The Mysteries of Modern Science

The Mysteries of Modern Science

Brian M. Stableford

LITTLEFIELD, ADAMS & CO.

Published 1980 by
LITTLEFIELD, ADAMS & CO.
by arrangement with Routledge & Kegan Paul Ltd.

Library of Congress Cataloging in Publication Data

Stableford, Brian M
     The mysteries of modern science.

     (Littlefield, Adams quality paperback; no. 360)
     Reprint of the 1977 ed. published by Routledge and Kegan Paul, London.
     Includes index.
     1. Science—History. 2. Science—Philosophy.
3. Science—Social aspects. I. Title.
Q125.S738   1980   509'.034   80-16332
ISBN 0-8226-0360-8

Printed in the United States of America

Contents

Introduction

Science is a match that man has just got alight. He thought he was in a room – in moments of devotion, a temple – and that his light would be reflected from and display walls inscribed with wonderful secrets and pillars carved with philosophical systems wrought into harmony. It is a curious sensation, now that the preliminary splutter is over and the flame burns up clear, to see his hands lit and just a glimpse of himself and the patch he stands on visible, and around him, in place of all that human comfort and beauty he anticipated – darkness still.

H. G. Wells, *The Rediscovery of the Unique*

One of the first men who thought that he had discovered the key to the riddle of the universe was Pythagoras. He found that if a string, when plucked, sounds a C, then a string half its length will also sound a C. So will a string twice as long. If the string is cut into three, then each third will sound a G. If it is cut into five, each fifth will sound an E. On the basis of this evidence, Pythagoras concluded that the harmony of nature was to be found in numerical ratios. He came to believe that if one understood the relationships between numbers one might understand the nature of the world.

It is interesting to look back and consider how Pythagoras used his discovery, and how it served him. He founded a school of philosophy; an élite corps which would protect and care for the discovery, and carry on the work of figuring out the universe. Then he went on to make the discovery which still bears his name, and which – ironically – shattered his dream. He discovered that the square on the hypotenuse of a right-angled triangle was equal to the sum of the squares on the other two sides. From this it followed that the diagonal of a square was related to the length of its side by the square root of two: a number which defied expression as a decent fraction. It could not be expressed as the ratio of two whole numbers and was therefore a terrible flaw in the numerical harmony of nature.

1

The reaction of the Pythagoreans was one of horror. They termed the square root of two an 'unspeakable number' and resolved to treat it as such. They swore never to disclose it, but the secret leaked out, as secrets always do. So much for the first answer to the riddle of the universe.

In a sense, the real achievement of Pythagoras was not the discovery of the magic of simple arithmetic, but the discovery that the universe is a peculiar place and cannot easily be fitted into a pattern of preconceptions, no matter how elegant and aesthetically compelling such a pattern might seem.

Ever since Pythagoras, men have been discovering keys by which they sought to unlock the mysteries of the universe. Some of the keys were quite useful, and many doorways to understanding have been unlocked. But behind each door, intrepid explorers in search of 'The Answer' have found only a limited space. In the modern age we have a vast edifice of understanding which goes by the name of science, which has many rooms whose doors have been unlocked (and where some rooms have been gathered together into suites, or even whole floors). Nevertheless, there remains a universe of mystery beyond doors still unlocked; and in addition, we now find some of the rooms so strangely decorated, and their keys so difficult to use, that we find as much mystery as we can cope with within the understanding we already have.

In a later age, Pythagoras's 'unspeakable number' was re-named an 'irrational number'. Mathematics came to terms with the square root of two, and with its cousin *pi*. (More than two thousand years passed, however, before it was finally *proved* that such numbers could not be expressed by the ratio of whole numbers.) Attempts to repair the flaws in nature introduced by these wicked terms continued into the twentieth century: a member of the State Legislature of Tennessee suggested making *pi* equal to three within the State by passing a law to that effect. Further to this point, an English clergyman, in one of those brilliantly silly letters which English clergymen are always sending to *The Times*, suggested passing a law making it equal to four – he felt that three was inadequate because it was not an even number. People have always been willing to defend their notions of the harmony of nature – a German philosopher, Vaihinger, has written a book on the tendency called *The Philosophy of 'As If'*.

There is no doubt that if the 'law of nature' had any justice about it, *pi* would have been set equal to three (or four) in the first place, and the square root of two would likewise have been decently clothed with arithmetical neatness. But there is, it seems, no justice at all in the law of nature. And even if the State Legislature of Tennessee had taken the first pioneering step in setting things to rights, one is inclined to doubt that the circle would have capitulated.

The simple fact is that reality is not so simple. The riddle of the universe is, like most riddles, a trick question, and the more answers we discover, the more tricks emerge to confound them.

It is popularly considered that in the twentieth century – known to some as the Age of Discovery and to others as the Age of Anxiety – science has come closer than ever before to a *true* understanding of the universe. In order to do so, it has not only had to build a philosophy which accepts irrational numbers as rational, but also one which accepts imaginary numbers (which are even more indecent than irrational ones) as real. The simple answers have, of necessity, been abandoned. The universe revealed by the discoveries of twentieth-century science is a universe of extreme peculiarity, and science has been forced to evolve a peculiar brand of understanding in order to cope with it.

Twentieth-century science has taken explorers well beyond the limits of their own perceptions, into the microcosm of the atom and the macrocosm of the expanding universe. It has revealed vistas in time within which the twentieth century itself is a negligible span. The revelation of these strange realms beyond perception has forced the scientist – for the first time – to seek keys to understanding which are not merely *as* strange as, but *more* strange than the world he sees around him. The Greek scientists, in the wake of Pythagoras, had to come to terms with the fact that measuring the world around them was not so easy. The twentieth-century scientists have had to come to terms with something much more difficult – the fact that the world which they can see is only a tiny fraction of what there is to be measured, and that the strangeness of the visible world is only one aspect of the strangeness of the whole. The limits of common sense have been surpassed.

Modern scientists do not, for the most part, copy Pythagoras in trying to conceal the dreadful complexity of nature behind a façade of assurance, though there are still echoes of the Pythagorean closed community in the scientific establishment. (No scientist is eager to admit that the idea which yesterday he believed to be the key to the universe is tomorrow's intellectual garbage.) Twentieth-century scientists concede what the Pythagoreans would not − that if the universe seems to be irrational then we must modify our rationality to accommodate it. If common sense is inadequate, then we must find some uncommon sense.

There are no more answers as easy as ABC or 12345. Modern science puts tremendous demands on the imagination, asking us to conjure up concepts for which we have no ready models in the world around us. There are no simple analogies which will allow us to picture an electron or black hole.

Modern physicists have discovered a strange kinship between the realms of reality which they are exploring and the bizarre imaginary worlds of Lewis Carroll which defy common sense so beautifully and so unrepentantly. Alice's Wonderland, it seems, may contain better models of certain aspects of reality than the narrow world shown to us by our senses. (Martin Gardner's *Annotated Alice* is an example of a modern mathematician's fascination with the imaginative constructs which sprang from the Rev. Dodgson's own mathematical ideas. George Gamow called his own attempt to simplify and popularise the concepts of modern physics *Mr Tompkins in Wonderland*. Sir James Jeans described the physics of Clerk Maxwell as an 'enchanted fairyland'. The 1974 series of the traditional Faraday Christmas Lectures at the Royal Institute, given by Eric Laithwaite, was titled 'The Engineer Through the Looking Glass'.) Today, Alice has become Everyman, trying to make nonsense into sense while, all the time, the world gets 'curiouser and curiouser'.

But Alice, like Pythagoras, revolted against the complexity of it all. Wonderland was written off as a dream, and she came back from the world beyond the looking glass. It is not so easy for us. We cannot really unlearn what we know, although there are considerable movements which attempt to do exactly that: trying desperately to find easier answers or a simple context in which to put it all. Somehow, we must learn to live with the uncomfortable revelations of twentieth-century science.

The mystification of science is something which has happened only recently − perhaps within the last hundred years. Early scientists were not so much seekers of absolute truth as seekers of easy answers to complicated questions. Pythagoras was strongly attracted to the idea that one could rationalise everything with the aid of simple numerical relationships not only because it seemed to fit the facts, but also because it seemed so eminently *orderly* and *sensible*. Scientists throughout history, in searching for the general rules which lie behind the chaos of singular events, have always earnestly believed that such rules must be elegant and reasonable. All scientific endeavour, in fact, is founded upon the assumption that behind the chaos of singular events there *are* unchanging, rational, coherent principles. Scientists used to believe that the answer to the riddle of the universe would, like the answers to all good riddles, be possessed of the quality which Shakespeare dubbed the soul of wit: brevity.

This faith which the scientist had in simple, comprehensive answers reached its peak in the nineteenth century, when it seemed – for a while – that answers to all possible questions were emerging from the darkness. Within the space of a few decades Darwin rationalised evolution, Dalton and Mendeleev rationalised chemistry and Pasteur rationalised medicine – a tripartite revolution comparable to that wrought by Newton in astronomy and physics two centuries earlier.

One of the men most inspired by the triumphs of nineteenth-century science was Ernst Haeckel, a German evolutionist who fancied himself something of a philosopher (as all inspired scientists seem to do). In 1899 Haeckel published a book entitled *The Riddle of the Universe* – a work which is most interesting in that its version of both the riddle and the answer typify the supreme self-confidence of the nineteenth-century rationalists.

Haeckel undertook as his task the clarification of seven 'world-enigmas' which had been put forward in 1880 by a French philosopher (not a scientist), Emil du Bois-Reymond, as the great mysteries of the universe.

These world-enigmas were:

1 The nature of matter and force.
2 The origin of motion.
3 The origin of life.
4 The orderly arrangement of nature.
5 The origin of simple sensation and consciousness.
6 Rational thought and the origin of the cognate faculty, speech.
7 The freedom of the will.

Three of these enigmas (1, 2 and 5) were, according to Bois-Reymond, transcendental and insoluble. Three others (3, 4 and 6) he considered soluble, but extremely difficult. Concerning the last, he was undecided.

Haeckel would have none of this. 1, 2 and 5, he argued, were perfectly simple questions, easily understood through the 'law of substance' – a combination of the law of conservation of matter and the law of conservation of energy which Haeckel represented as the fundamental law of physics. Enigmas 3, 4 and 6, he claimed, were easily understood with the aid of Darwin's theory of evolution. With enigma number 7 even Haeckel might have had difficulty, but – like Alexander confronted by the Gordian knot – he was unconfounded. 'The freedom of the will', he wrote, 'is not an object for critical scientific enquiry at all, for it is a pure dogma, based on an illusion, and has no real existence.'

Haeckel's standpoint on these matters was perhaps a little extreme even for the nineteenth century, but his casual assumption that a little knowledge is an all-embracing thing has never been uncommon among scientists. (Perhaps we should not hold it against Haeckel, as his free will, being illusory, seems to have given him little choice ....) If scientists today are more modest, it is probably because they recall the absurd spectacle of their nineteenth-century counterparts, whose all-inclusive knowledge of the universe was proved totally inadequate in a few decades, rather than because they are much more aware of the deficiencies of their own understanding.

A contemporary of Haeckel's, the chemist Marcellin Berthelot, is today better remembered for an off-hand remark he made in 1887, to the effect that 'From now on, there is no mystery about science', than for his actual contributions to the sum of knowledge. He was not only wrong, but wrong at precisely the wrong moment in history, for science became progressively more mysterious thereafter. Haeckel and Berthelot were seduced by the charisma of the brilliant simplifications wrought in scientific theory by Dalton and Darwin. Two centuries had found no fault with Newton's achievements – why should these be undermined or re-appraised any more rapidly? They could not know that even the classical science of Newton would be out of fashion before too many years had passed.

The development of chemistry perhaps provides the best example of the pattern of simplification and re-complication. The nineteenth-century chemists inherited a confused science which had never quite abandoned the attitudes and methods of the alchemists. There seemed little rational pattern in the endless catalogue of substances. Dalton believed, however, that there was a relatively small number of *basic* substances – elements – and that all other substances consisted of compound molecules made up of atoms of the elements combined in specific fixed proportions. This made possible the *formularisation* of chemistry, the establishment of chemical equations every bit as 'correct' as mathematical equations. The same atoms always combined to make the same molecules the same way.

In 1869 Mendeleev introduced a further simplification when he grouped the elements into a rational hierarchy – the periodic table. Mendeleev discovered a *system* in the variety of the elements, and showed how they conformed to a pattern. He not only provided for elements as yet undiscovered but set a limit on the possibilities which still existed. By this time the use of the spectroscope in identifying elements by means of the light emitted or absorbed by their incandescent gases had enabled Bunsen and Kirchhoff to begin ascertaining the composition of the stars, establishing that even they were composed of ordinary, mundane elements which fitted into the system.

It is easy to see how this pattern of discovery could affirm faith in the idea that nature was fairly simple once its principles were sorted out. The universe, it seemed, had less than a hundred elements, which could be sorted out into behavioural classes, and whose various combinations could be plotted and rationalised. Even the organic compounds, the stuff of life, which seemed, at first, to belong to an entirely separate order of nature, were recruited into the fold in 1828 when Wöhler first synthesised urea in the laboratory, and – later – when Kekulé found the key to identifying the 'structural formulae' of complex organic compounds.

The echoes of the Pythagorean inspiration are also clear in the *nature* of the discovered system: in Mendeleev's classification the elements were primarily distinguished by *atomic number*, and the vast array of possible compounds were systematised by identification of the simple whole-number ratios in which atoms could combine. Once again, it seemed that the common-sense rules of arithmetic were adequate to an understanding of the working of the world.

Ultimately, of course, the atomic number of an element proved to be something *real* – the number of protons in the nucleus. But by the time this was discovered, the beautiful simplicity of the atomic theory was a thing of the past. Elements were no longer elementary, but made up of smaller constituents whose behaviour-patterns were often very odd indeed. At first, it seemed that beneath the simple array of the ninety-odd elements lay an even simpler array of subatomic particles, three in number, combined in simple numerical ratios, but it proved not to be so. There proved, ultimately, to be more subatomic particles than elements, and the logic of their behaviour defied simple arithmetic entirely. In addition, the simple pattern of the elements had been considerably disturbed by new discoveries. A whole series of elements – the rare earths – were squeezed into a single place in the periodic table. Some elements, it seemed, had a disconcerting habit of changing spontaneously into other elements. Virtually all elements had isotopes – that is to say, more than one kind of atom.

The same kind of pattern can be seen in the historical development of all scientific disciplines. 'Progress' appears to be a very sporadic process. When a number of basic principles are suddenly illuminated (evoking the well-known Eureka effect) whole conglomerations of confused data suddenly fall into place and become orderly. The whole picture becomes clearer – much as the world does when a short-sighted person gets a new pair of glasses. However, the simplified picture immediately begins to accumulate new galaxies of confused data which arise as exceptions and modifications, so that the simple becomes complex again, and new answers have constantly to be found for new problems. (If only this pattern of change were itself systematic, it would not be so bad, but it is not. Minor inconsistencies and exceptions have a tendency to be shunted aside, forgotten or ignored while scientists cling to their elegant – and inadequate – answers, and are hardly ever evoked or noticed until a new Eureka effect sweeps them into prominence along with its victorious explanations. Thus science always *seems* to have the answer to the riddle all worked out – and then along comes a whole new perspective.)

In a sense, it was *because* the nineteenth century saw such devastating simplifications, and brought such magnificent order to so wide a spectrum of knowledge, that the re-complication and re-mystification of the twentieth century has been equally devastating and diverse. It was because the enlightenment was so wonderful that we have been able to get such a flying start into the unknown once again. If the nineteenth-century scientists had not come to see so far, so clearly – and tried hard to make others see with their inspired eyes – then their intellectual heirs might not have discovered all the blurred patches and obscure corners of their world-vision so quickly, or attacked them so fiercely. Had Haeckel and his contemporaries not been so very *public* in their self-acclaim (had they, in fact, emulated the Pythagoreans in trying to keep science esoteric) the mysterious universe we live in today might still be lurking beyond our conceptual horizons.

There are undoubtedly some who would prefer it so. Neither the discovery of the neutrino nor the search for a neutron star has added greatly to the general morale of mankind, and the revelation of the possible applications of atomic chain reactions have had rather a deleterious effect in this respect. But it is now too late to wonder whether Pythagoras might not have been right, and that mankind would have lived a happier, more carefree existence if the dread secret of the square root of two had been buried forever. We must, instead, adapt ourselves to the world we have discovered and created.

It is unfortunate that twentieth-century science remains, for the most part, outside the scope of the standardised education mass-produced for the majority. It is, of course, impossible for such a standardised product to be comprehensive, and it is understandable that it should concentrate on those things deemed 'essential'. I wonder, however, whether our ideas about what is essential are not out of date. It is easier to teach simple answers than complex ones, but to opt for simplicity at the expense of reality may well be a false economy.

The great majority of the victims of modern-day education never come into contact with twentieth-century science – it is left to a relatively tiny group of specialists. The rest may not encounter science at all, but if and when they do they are offered a dilute cocktail of nineteenth-century science. It is not really the *factual* inadequacy of such an encounter which is important, but the inadequacy of *perspective*. The out-of-date science taught in modern schools carries with it the out-of-date attitudes which came to the fore when common sense was king. Thus scientific education in school is not merely inadequate, but misleading.

There are problems of perspective, too, in the rarefied regions of higher education where initiation into selected mysteries of modern science actually takes place. By that time, educational specialisation has set in, and anything the consumer gains in depth he is apt to lose in breadth. The greater the knowledge one acquires through the educational system, the narrower becomes the context in which one is able to apply it.

There is a certain irony in the fact that though the power of the world's stockpiled hydrogen bombs is sufficient to annihilate its entire population several times over, not one in a million of those threatened can begin to understand how such a device can exist. Similarly, though television has become the new opium of the masses in developed countries, the millions who live their lives second-hand *via* the softly glowing screen have no idea how or why it glows. This problem goes deeper than mere ignorance − which is supposedly voluntary in a literate society. The problem is one of adaptation to a whole new world of *ideas* − a new series of perspectives which go beyond the limits of those retailed in the economy-size educational packet. It is not enough to know the mechanics of the TV set and the hydrogen bomb − one has to get the feel of the wide scientific context to which these objects belong. Twentieth-century science is remaking the world from top to bottom, and it is not only the view from the ivory towers which looks new, but the scenery in the living room. We are strangers in our own artificial environments.

The popularisation of science began in the nineteenth century with the advent of near-universal literacy, and there is still considerable activity in the field today. Men like Isaac Asimov have done excellent work in making the factual component of twentieth-century science available to the layman, translating jargon and clarifying, step-by-step (historically and logically) how the discoveries of modern science were made and what their implications are.

At the same time, however, there is considerable activity in popularising alternatives to science. The market in simple answers to the riddle of the universe is definitely a seller's market and trade has probably never been so good. Many varieties of 'alternative knowledge' are current and apparently gaining ground: Velikovskianism, astrology, scientology, Oriental mysticism, etc., etc. The success of these alternatives (irrespective of any fidelity to truth which they may contain) is testimony to the failure of modern education to supply the intellectual armaments needed for survival in the contemporary world.

It might be argued that the failure is the failure of science itself to answer the questions − the riddles − which people want answered, and there is probably some truth in this. The answers which modern science might venture in reply to Bois-Reymond's seven world-enigmas are no less bleak than the answers provided by Haeckel and certainly a good deal less self-assured. It seems to me, however, that if it is science *per se* which has failed to supply a demand or a need, then it is difficult to explain why so many of the new alternatives choose to mimic it so closely, carefully husbanding every scrap of evidence, counterfeiting jargon and everywhere appealing to the suspect logic of common sense. It appears that people *do* have faith, of a kind, in scientific methods and perspectives, but that it is only a relic of nineteenth-century confidence in sweeping, all-embracing theories of great simplicity. This is the kind of science which people feel at home with.

In my opinion, popularisation of the facts of modern science is only half the job that needs to be done. It is also necessary – though somewhat more difficult – to popularise the *imaginative context* of twentieth-century science: its *mystique*. We must learn to feel at home with the ideas of modern science. We must adjust our outlook to a new intellectual climate.

I have not set out, in this book, to *explain* the mysteries of modern science, but rather to *explore* them. This is not a specialist text but a general survey of the kinds of idea which have surfaced as a result of twentieth-century advances in science. It is concerned with implications (imaginative and social) rather than with details of fact and theory or with mathematical exposition. It is, I hope, an elementary guide-book to the conceptual universe of modern science. Like the kind of tourist guide which attempts to introduce you to the heart and soul of a foreign country in the space of a ten-day package holiday, it has its limitations – I will do the best I can within them.

I have divided the exploratory programme into three parts. The first investigates the infinite universe revealed by physical science, and the imaginative horizons discovered therein. The second investigates one very thin layer of the universe and one of its more peculiar phenomena: life. The third looks at the invasion of everyday life by the products and perspectives of science.

Part One
The Surrealistic
Cosmos

Our only hope of understanding the universe is to look at it from as many points of view as possible. This is one of the reasons why the data of the mystical consciousness can usefully supplement those of the mind in its normal state. Now, my own suspicion is that the universe is not only queerer than we suppose, but queerer than we *can* suppose. I have read and heard many attempts at a systematic account of it, from materialism and theosophy to the Christian system or that of Kant, and I have always felt that they were much too simple. I suspect that there are more things in heaven and earth than are dreamed of, or can be dreamed of, in any philosophy. That is the reason why I have no philosophy myself, and must be my excuse for dreaming.

J. B. S. Haldane, *Possible Worlds*

1   The Limits of Common Sense

I recently came across an article which began with the words: 'The idea of relativity is really fairly simple.' I assume that the writer was attempting to reassure his readers rather than to patronise them, but I do not suppose that he succeeded. The attitude represented by this kind of statement is quite common among scientists — it is the sort of thing one might imagine Haeckel saying to Bois-Reymond. It may be expanded thus: 'Look, all this is perfectly simple and makes excellent sense to *me*. If it doesn't make sense to you then you must be stupid.' There is an old anecdote concerning a debate between the French Encyclopédist and atheist Denis Diderot and the Swiss mathematician and confirmed Christian Leonhard Euler about the existence of God. Euler said: '$(a + b^n)/n = x$, therefore God exists' and Diderot could find absolutely no way to refute the statement.

It is quite easy for a scientist to concoct mathematical logic and literally blind his opponent with science, as Euler did to Diderot. It is not simply that what is obvious to one man is not obvious to another, but that the kind of argument that one man recognises and accepts is not necessarily the kind of argument which makes sense to someone else, irrespective of stupidity. This applies just as much to honest scientific discourse as to Euler's intellectual trickery.

Perhaps, as the article-writer claimed, the idea of relativity is really fairly simple. But it is not an easy one to grasp, and it is certainly not easy (or simple) to extrapolate the basic idea and discover its logical consequences. It requires a good deal of intellectual effort to surrender the notions of common sense which serve so well in the everyday world in order to adopt rather peculiar notions which, though they may confer a certain intellectual *savoir faire* on their owner, are never likely to come in useful in the routine of daily life.

I, personally, remain convinced that it is a *worthwhile* exercise to try to acquire uncommon sense, in order that one may come to terms intellectually with the infinite and the infinitesimal, but this does not imply that I find the ideas of modern physics simple or easy to command. Therefore, I begin with no promises about how easy to understand everything I am going to write about in these first chapters really is, nor will I covertly imply that anyone who remains unconvinced is a moron. I will avoid mathematical language and mathematical logic as much as possible, as these esoteric tools may confound even men like Diderot.

An introduction to twentieth-century science cannot, alas, begin at the beginning, because the beginnings are lurking in the shadows of the distant past – perhaps even the pre-Pythagorean past. Perhaps, then, it is best to begin with something that is familiar, if only by repute: the myth of Albert Einstein.

Einstein's name has passed into the language of everyday life, a word whose connotations are immediately recognisable. His name has joined those of Samson and Napoleon, Sherlock Holmes and Tarzan as the symbolisation of some particular attribute. Einstein is a symbol for intellectual, theoretical genius (not to be confused with inventive, practical genius, whose figurehead is Edison). The magic formula which bears testimony to Einstein's genius is $E = mc^2$ ($E = mc^2$, therefore Einstein is a genius, as Euler might have put it).

What this equation means is that mass and energy (m and E) are basically the same kind of stuff, and that the rate of exchange by which one may be converted into the other is determined by the constant c^2 (c is the velocity of light in a vacuum).

The heart of the Einsteinian revolution in physics is the magic of c – the specialness of the velocity of light in a vacuum. Let us, then, begin by trying to see how it was that the nature and properties of light came to be so crucial in determining the way modern science looks at the world.

Newton, who wrote a treatise on light in the seventeenth century, considered that light consisted of tiny corpuscles which shot through the ether and stirred up little waves (much as a fish moving through water stirs up waves as it moves). He knew that there were certain respects in which light behaved like a particle – in reflection, for example, when it bounces back from a shiny surface – and others in which it behaved like a wave (as when two beams of light interfere with one another). His 'fluttering particles' were an attempt to reconcile the two sets of properties. His contemporary Huygens, however, saw no necessity to suppose that there might be corpuscles of light at all, and thought that light was simply a vibration in the ether, just as sound is a vibration in air. Sound waves, of course, can be reflected (as echoes) without there being any actual entity to 'bounce'.

Science has always held a great deal of respect for a principle called Occam's razor, a dictum dating back to the fourteenth century which says that in seeking explanations one should not invent more than is strictly necessary. According to the scientists who came after Huygens and Newton, it was necessary to see light as a wave of some kind, but to invent a particle to cause the wave was an unnecessary embellishment. It was only *necessary* to envisage a disturbance moving through the ether, and that was how scientists during the next two centuries did envisage light.

All new discoveries concerning light made during the years which followed tended to confirm this view – they could all be explained in terms of wave phenomena. The theory needed no elaboration at all. And then, in the nineteenth century, the theory was suddenly extended so that as well as explaining light it explained many other physical phenomena. Just as Dalton unified chemistry with his atomic theory, so James Clerk Maxwell began to unify physics with a theory of waves.

Maxwell exemplifies the spirit of modern science in several ways. He is perhaps best remembered today in the phrase 'Maxwell's demon', which refers to another change in attitude which he brought about. He brought an entirely new perspective to the kinetic theory of gases. The temperature of a gas depends on the activity of its molecules, and Maxwell was the first to point out that it is a measure of the *average* activity of the molecules, and that individual molecules may be going much faster or slower, and are in any case always transferring their energy to one another in collisions. The first law of thermodynamics is that heat always passes from a hot system to a cooler one, but Maxwell showed that this was only the overall result of a large number of individual events whereby molecules transfer their energy. On balance, over a period of time, the slower molecules tend to be speeded up by interaction with faster ones, and the faster ones slowed down, but any individual interaction could go the other way.

Maxwell's demon was a hypothetical being who might sit at a

small doorway connecting a hot gas with a cool one. This demon would only let the fastest molecules from the cool gas pass through the doorway, turning the slower ones back. At the same time he would only let the slowest molecules of the hot gas pass through, turning the faster ones back. By this means, he would ensure that the hot gas got hotter and the cool gas cooler.

There are two important things to note about this piece of theorising – the first is that the law of thermodynamics was shown to be a result of the statistical aggregation of a large number of events rather than an inviolable principle ruling the world with an iron hand. The second is Maxwell's way of dramatising this point with the example of the demon – an imaginary being performing an imaginary task. Maxwell's demon is a kind of thought experiment – an imaginary experiment incapable of actualisation, but capable of making a point in much the same way as a real experiment does. Although the demon is unreal, it is intellectually satisfying, because with its aid we can begin to see that although the law of thermodynamics always works out *in practice*, it could, in fact, be subverted by an extremely unlikely combination of chance happenings – it is not a law so much as a statistical prediction.

These things are important because in the twentieth century virtually all physical laws have been shown to be, like the laws of thermodynamics, statistical predictions and not inviolable principles; and also because virtually all the most important experiments of twentieth-century science have been thought experiments and not real ones: thus has scientific work been taken to a large extent out of the laboratory and into the imagination.

Maxwell's contribution to the physics of light was to tie it in with the physics of electromagnetism. Prior to Maxwell, the way in which a magnet attracted a piece of iron had been understood in terms of 'tubes of force'. Physicists had imagined the magnet equipped with invisible tentacles which reached out and grabbed the unmagnetised iron. This is a common-sense analogy – one can see in one's mind how the magnet works, thanks to this idea.

Maxwell, however, saw magnetism in quite a different way. He envisaged a force-field which existed all around the magnet, so any piece of iron within that field would become subject to a force whose direction and magnitude could be determined mathematically. This analogy is not so much a visual analogy as an abstract mathematical one: the force-field is not so much a thing as a set of equations. But Maxwell tried to make his force-field an entity, to allow it to be visualised, by supposing the existence of an 'electromagnetic ether'. In this electromagnetic ether there were electrical and magnetic *tensions* – there was energy there just as there was energy in a taut string or a twisted rubber band, and under the right circumstances this energy could become force – electrical or magnetic. When a wire moved through a magnetic field, electricity was generated, and when magnetised iron was placed in the field, it became magnetised and was attracted to the magnet. (And, just as the notes sounded by plucked strings are mathematically predictable, so the behaviour of wires subject to 'magnetic plucking' could be determined mathematically.)

All this was simply a shift in perspective. But it allowed Maxwell to come up with a very novel idea. If there was an electromagnetic ether full of electromagnetic tensions, then there was a theoretical possibility of electromagnetic *waves*. Occam's razor came to hand again – why have two sorts of ether and two sorts of wave, when one might be enough?

And so, Maxwell suggested, might it not be likely that light actually consisted of electromagnetic waves in the ether? All this he had accomplished by thought experiment – by trying to *imagine*, inside his head, the kind of thing which was going on to produce the physical phenomena he was studying. It was in simply replacing one analogy by another – one way of thinking by another – that Maxwell made this discovery.

The inspiration, like Dalton's and Darwin's, convinced the nineteenth-century scientists that they were literally getting it all together. Maxwell's insight was so beautifully elegant and simple that it seemed that it must be true. But thought experiments are still only thought experiments, and it remained to be demonstrated that there actually *were* electromagnetic waves in the ether.

In 1878 David Hughes demonstrated before the Royal Society that signals generated by an electrical spark could be picked up by a microphone in circuit with a battery and a telephone a quarter of a mile away, and claimed that electromagnetic waves were responsible for the transmission. The Royal Society was unimpressed and considered that the existence of waves was unproven. Nine years later Hertz made a more convincing show, and demonstrated conclusively that radio waves generated electrically could be picked up at a distance. Hertz had been interested purely and simply in proving Maxwell's theory, and having done so he advised that research into electromagnetic waves should be discontinued as it was of no conceivable practical value; thus showing that scientific expertise is not necessarily coupled with a practical imagination. It was left to Marconi to develop wireless telegraphy and pave the way for radio.

Even in Newton's day it had been accepted that the different colours seen in spectra represented different frequencies of vibration. (The fact that we recognise seven colours in the spectrum − red, orange, yellow, green, blue, indigo and violet − is, in fact, due to Newton's fondness for the number seven: he included orange and indigo as separate colours although they do not really show up distinctly.) It was therefore obvious that Hertz's invisible waves differed from light in terms of the frequency of their vibration. Thus it became clear that the spectrum of coloured light might well be only a tiny fraction of a much broader spectrum in which all conceivable frequencies of vibration were represented. All kinds of invisible rays might exist in parallel with rays of light.

And so it transpired. Less than a decade after Hertz first demonstrated radio waves, Röntgen discovered X-rays. Hertz's discovery had met with relative indifference, but X-rays captured the public imagination, their possible practical applications being realised at once. The use of X-ray photographs in the diagnosis of bone diseases, the location of fractures and the investigation of tuberculosis caught on very quickly, and X-raying even became a popular curiosity in stage magic. The twin concepts of rays to heal disease and rays to kill − along with other rays to turn lead into gold, armour into dust, and to provide power for all manner of miraculous machinery − became a prominent feature in imaginative literature.

The discovery of the radioactivity of certain elements followed very quickly, and when Rutherford and Soddy formulated a theory of radioactive decay in 1903 they identified three different types of radiation, one of which was an electromagnetic wave they christened the gamma-ray.

The entire electromagnetic spectrum was soon categorised. The velocity of the various rays seemed always to be the same: c. This meant that the frequency of any particular wave (i.e. the number of pulses per second) was inversely related to its wavelength (i.e.

the length of the pulse). Visible light filled a very small range of frequencies at about 10^{15} (a thousand million million, or 1,000,000,000,000,000) cycles per second. Radiation of higher frequency was classified into ultra-violet, X-rays and gamma-rays. Radiation of shorter frequency but longer wavelength was classified into infra-red, microwaves, and radio waves.

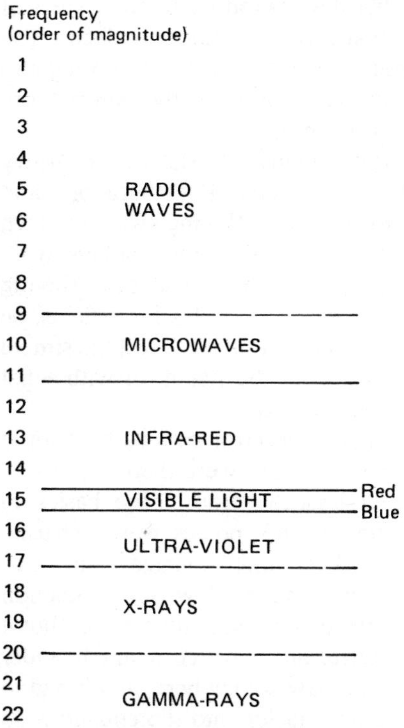

Frequency
(order of magnitude)

1	
2	
3	
4	
5	RADIO
6	WAVES
7	
8	
9	— — — — — — — —
10	MICROWAVES
11	————————
12	
13	INFRA-RED
14	
15	VISIBLE LIGHT — Red
16	— Blue
17	ULTRA-VIOLET — — —
18	
19	X-RAYS
20	— — — — — — — —
21	GAMMA-RAYS
22	

Figure 1   The Electromagnetic Spectrum

Figure 1 shows a diagram of the spectrum. The scale on the left is the *order of magnitude* of the frequency in cycles per second (thus $1 = 10^1 = 10$; $2 = 10^2 = 100$; $3 = 10^3 = 1,000$, etc.). The wavelength corresponding to 1 cycle per second is 3×10^8 metres (300,000 km) and the wavelength corresponding to 10^{20} cycles per second is 3×10^{-12} metres (0·000000003 mm). Unlike a linear scale, this kind of scale is open-ended, the sequence of the negative numbers signifying ever smaller fractions. The spectrum thus has no 'ends', but extends into the infinite and the infinitesimal.

The discovery of the full range of the electromagnetic spectrum and the understanding of its nature proved extremely useful in many practical applications. Edison, for instance, had already discovered a method of turning electricity into visible light via the electric incandescent bulb. Other devices radiate in other areas of the spectrum: a fluoroscope makes use of X-rays, an electric fire radiates mostly in the infra-red, and an ultra-violet lamp radiates the high frequencies which stimulate tanning in the skin without the infra-red frequencies which burn it.

But while the practical applications of Maxwell's theory became a veritable flood, its theoretical elaboration and ramification did not proceed so smoothly. Having switched from a physical analogy to one which was primarily mathematical, Maxwell's ideas were at the mercy of mathematical logic. Having found the fundamental equations determining his force-fields, he had to live with the corollaries of those equations, and the simple fact was that some of those corollaries did not fit in with what actually happened out there in the world.

Maxwell's equations describing the electromagnetic field could be used to make accurate predictions and were therefore 'true'. But the equations worked out logically from those basics to explain electromagnetic radiation by hot bodies were *not* true. According to the equations, whose mathematical logic was impeccable, the hot bodies were doing it all wrong. Radiation *ought* to escape in a quick, catastrophic burst of ultra-violet, but it did not. It was much more sedate and covered a much wider range of frequencies. There was a flaw somewhere, and it had to be overcome by introducing a new factor into the equations so that they behaved properly.

In order to iron out this flaw in Maxwell's mathematical theory, Max Planck introduced the idea of the quantum. According to Planck, energy had to be emitted (or absorbed) in measured bursts – the energy of the wave had to be packaged in predetermined quantities (in much the same way that cigarettes are packaged in tens or twenties, not thirteens, or that currency notes package £1 or £5 but not £3·45). The amount of energy in any particular package depended, said Planck, on the frequency of the wave, and was related to it by a constant factor called Planck's constant. (Like c, Planck's constant – h – is one of the universal fundamentals.)

The idea of electromagnetic quanta was rather alien to the attitude that light was simply a disturbance in the ether. It was almost reminiscent, in fact, of Newton's 'fluttering particle' in that it implied that a light-wave had to be a specifically structured disturbance, almost an entity in its own right. The whole point of the Maxwellian revolution, however, was that it did away with clumsy hypothetical entities altogether and replaced them with streamlined mathematical concepts. Physicists were definitely *not* keen to resurrect Newton's corpuscles. Planck had, in any case, only said that light had to be emitted or absorbed in measured amounts – he had not envisaged it being quantified while it was propagated in the ether. The man who came forward to suggest that light energy was *always* packaged in quanta, while it was travelling as well as when it was departing or arriving, and thus re-introduced the 'atom of light' – the *photon* – was Einstein.

Einstein was a clerk in the Swiss patent office at Berne. He was hoping to become a teacher, but spent his evenings studying higher mathematics – he was particularly interested in the 'imaginary geometry' of Lobatchevsky and Riemann. All Einstein's discoveries were made inside his head, in his imagination. Not only did he never carry out an experiment, but he was not even an observer of phenomena. He dealt entirely in ideas.

The first major idea which he contributed to physics was the photon, and the justification of its proposal was that it explained a minor anomaly detected by Hertz which had been dubbed the 'photoelectric effect'.

Hertz had noted that when ultra-violet light shone on a metal surface, electrons (small, electrically charged particles – the 'units' of electricity and the first subatomic particles to be identified) were emitted. According to Maxwell's theory, when the intensity of the light was increased the energy (and thus the velocity) of the emitted electrons ought to increase. In fact, the velocity of the electrons stayed the same, but more were emitted. In order to obtain electrons of greater energy, the *frequency* of the light had to be increased.

Einstein pictured light as consisting of tiny arrows, each one containing a quantum of energy. According to this model, increasing the intensity of the light was only sending out more arrows of similar energy – which could knock out more electrons but could not kick them any harder to make them go faster. Increasing the frequency of the radiation, however, bound up a larger quantum in each arrow, and more energy was thus passed on to the electrons.

For this explanation (not for the theory of relativity) Einstein received the Nobel prize.

In a sense, Einstein's discovery of the photon was a reversion to pre-Maxwellian thinking about light. Maxwell had been the first man to abandon the search for scientific truth in terms of physical illustration and opt, instead, for the symbolic expressions of pure mathematics. In his wake, all the classical concepts of physics whose meaning seemed solid began to dissolve, by degrees, into mathematical abstraction. A school of philosophy, spearheaded by Ernst Mach, had emerged determined to carry through Maxwell's initiative and banish from physics all its 'imaginary entities' (matter, space, time, etc.) and make it a mathematically descriptive rather than a materially interpretative science. When Lorentz sought to replace the homogeneous 'electric fluid' with units called electrons they opposed him, and when Einstein sought to cut light-waves up into photons they opposed him. In each case, however, the particulate model proved more useful than the continuum model in providing explanations.

Thus, Maxwell had replaced clumsy illustrative analogies by equations, and now a new generation of physicists was developing *from* Maxwell's equations a whole new set of hypothetical entities. The essential difference between the new entities and the old was that they could no longer be visualised, for they were not the products of common sense but the synthetic specifications of sophisticated mathematics. Einstein's photon was by no means the same kind of thing as Newton's fluttering particle, although it seemed strangely to echo it. It was put together backwards − while Newton had attempted to graft mathematical properties onto a solid, real *thing*, Einstein had started with a mathematical abstraction and had complicated it with the property of discontinuity. He had *not* made it back into a thing despite his attempt to make it more readily understandable by comparing it to an arrow.

It is convenient now to return to Occam's razor. Newton's description of light had supposed a particle, a wave and an ether. Huygens, with Occam's razor gleaming in his hand, had claimed that only the wave and the ether were necessary to explain the facts. Maxwell had elaborated this idea, uniting the luminiferous ether necessary to carry light-waves with the electromagnetic ether he invented to carry electromagnetic waves.

But the ether had only been necessary in the first place in order that the wave could be visualised. In making the wave a mathematical abstraction rather than a real thing like a ripple in water or a vibration in the air, Maxwell had really dispensed with the necessity to suppose an ether. According to Occam's razor it ought to have been thrown out forthwith. Mach and his followers wanted to do just that. Others, convinced not only of the rightness, but also of the firmness, of nineteenth-century science, did not. Haeckel, for instance, wrote the following in *The Riddle of the Universe*:

> The existence of ether as a real element is a *positive fact*, and has been known as such for the last twelve years. We sometimes read even today that ether is a 'pure hypothesis'; this erroneous assertion comes not only from uninformed philosophers and 'popular' writers, but even from certain 'prudent and exact physicists'. But there would be just as much reason to deny the existence of ponderable matter.

The 'uninformed philosophers' referred to by Haeckel are, of course, Mach and his followers. But what was it that led to his confident assertion that the ether had been proved real for twelve years (i.e. since 1887)? He was, in fact, referring to an experiment carried out in that year by Michelson and Morley, and which has since come to occupy a most curious place in the history of science.

Michelson and Morley set out to measure the velocity of the Earth in the ether, and they proposed to do this by measuring the velocity of light in two different directions – one, in the direction of the Earth's motion, and the other, at right angles to it. They figured that in the first instance they would measure the velocity of light plus the velocity of the Earth, and in the second they would measure the velocity of light only. The difference between the two would be the velocity of the Earth in the ether.

This experiment is now held up as a classic case of a negative result disproving a theory, because we now know, or at least believe, that there is no ether and that the velocity of light in a vacuum is always the same. In actual fact, however, the Michelson-Morley experiment was decidedly ambiguous. They did find a difference between their two measurements, but it was nothing like as large as they expected, being considerably less than the known velocity of the Earth in its orbit. The result was open to interpretation in several ways: Haeckel was convinced that the discrepancy was enough to prove that the ether existed, others thought it probable that it could be written off as experimental error.

Nowadays, with the benefit of hindsight, many historians of science state quite definitely that Michelson and Morley found no significant difference. Curiously, however, the Michelson-Morley experiment was repeated thousands of times by D. C. Miller, using better apparatus – and Miller found exactly the same ambiguity. The discrepancy was not large, but it was there.

The reason why the Michelson-Morley experiment has become such a classic is that it is supposed to have paved the way for the theory of relativity – providing experimental evidence on which Einstein built his ideas (in the same way that the photoelectric effect required the invention of the photon to explain it). In actual fact, this is not so. The Michelson-Morley experiment did not figure at all in Einstein's thinking. It is significant that it did not, because it points to one of the chief hypocrisies of modern science. Science is represented by its practitioners as being in intimate touch with reality, with hypotheses emerging to explain observations, and experiments designed to test those hypotheses. This is the ideal blueprint for scientific progress. In fact, twentieth-century science is not like that at all: most of its hypotheses have arisen not to account for observations but to enhance the *aesthetic* qualities of equations and concepts, and most of its experiments have been designed to confirm rather than to test hypotheses (the difference is subtle, but it is important; ambiguous results, like those of the Michelson-Morley experiment, tend to become very *un*ambiguous in the eyes of the theorist who needs a particular result to confirm his ideas).

The theory of relativity, then, did not emerge from an attempt to explain observations made by earlier scientists. It was pure thought experiment – perhaps the purest thought experiment of them all. This is not to say that the idea sprang fully fledged into Einstein's head from nowhere, but its intellectual ancestors were themselves abstract concepts: the philosophy of Mach and the mathematics of Riemann.

Again, it is convenient to look back at what Newton thought.

Newton's discovery of the laws of motion and the principle of gravity had allowed him to rationalise the behaviour of the planets in their orbits. It is popularly said that he 'turned the universe into a great clockwork mechanism'. It is difficult now to appreciate the magnitude of this change of perspective: the transformation of the universe from a divine creation to a conglomeration of objects whirling around without purpose, simply obeying a vulgar natural principle: the law of gravity.

In turning the universe into a machine Newton not only reified it but supplied it with a whole series of absolutes within which its behaviour was regulated with the utmost rigidity. The universe-machine was situated in 'absolute space', which was at rest, and in which everything moved. It operated in 'absolute time' – everywhere shared the same ever-changing moment called the present. Both of these ideas seemed eminently sensible – they were, in fact, the epitome of common sense. The ideas of 'where' and 'when' are quite clear to us, and we know exactly what they mean.

However, when Mach came along with his declared intention of sorting all the convenient fictions out of physics and showing them up for what they were, absolute time and absolute space came in for some harsh criticism. There could be no possible proof that there was an absolute spatial framework 'at rest' while objects within it moved, nor could there be any proof of absolute time. Mach therefore contended that these concepts were without meaning. They were inventions of the human imagination, and were not real.

At the same time as Mach was attacking Newton, mathematicians were becoming dissatisfied with Euclid. Euclid had laid the foundations of the everyday, common-sense geometry – the plane geometry in which the angles of a triangle add up to 180°, parallel lines never meet and the square on the hypotenuse is equal to the sum of the squares on the other two sides of a right-angled triangle. That geometry had always served perfectly well for all practical purposes, but a number of nineteenth-century mathematicians, led by Lobatchevsky, became concerned to point out that plane geometry was, in fact, only one of a number of geometries: a special case of something much more general. Lobatchevsky became interested in what he called 'imaginary geometry' – working out the geometry implied by different assumptions, as, for instance, the assumption that parallel lines *did* meet. It became clear that all these geometries were logically self-consistent – they all depended, just as Euclid's did, on the assumptions one started with. And, just as there was no way to prove the existence of Newton's absolute space, so there was no way to prove the truthfulness of Euclid's postulates. They were convenient because they worked, but there was no reason to assume that they represented any kind of ultimate truth just because they made sense.

Other mathematicians, notably Riemann, became interested in Lobatchevsky's imaginary geometry, and particularly the geometry of four-dimensional space. We, of course, perceive space as three-dimensional, but that does not mean that space itself is three-dimensional. There is a classic work called *Flatland* (written by Edwin Abbott, a Shakespearian scholar whose hobby was higher mathematics, under the pseudonym 'A Square') which describes a two-dimensional universe existing within our own as a flat plane, and points out the difficulty a flatlander would have in trying to imagine the third dimension which exists outside his experience.

Imaginary geometry did not exactly catch on in a big way – and Lobatchevsky was, in fact, thought in certain quarters to be mad. Similarly, Mach's ideas, which seemed to be challenging the brilliant achievements of nineteenth-century science in banishing all mysteries and producing absolute truth, were virulently attacked by many scientists.

Einstein, however, not only adopted the ideas of Mach and Riemann, but went further than either. Where Mach had argued that absolute space and absolute time were meaningless, Einstein decided that they were actually wrong. Riemann had developed four-dimensional geometry as an exercise in abstract mathematics, but Einstein concluded that in actual fact the universe *was* four-dimensional and its actual geometry was Riemannian.

The thought experiment which convinced him of all this was really quite trivial. According to Maxwell, light consisted of electromagnetic waves travelling at about 3×10^8 metres per second. In that case, someone who was also travelling at 3×10^8 metres per second in absolute space would see the ray of light as a standing wave, like a vibrating string. In other words, to someone travelling at light-speed, the universe would appear quite different, and very strange. This seemed to Einstein to be silly. It offended his sensibilities, and was unaesthetic. It seemed to him that the universe ought to look the same no matter what one's actual standpoint was.

He concluded, therefore, that the idea of anyone travelling at light-speed relative to 'absolute space' was, in fact, nonsense.

There was, he decided, no such thing as absolute space, and all motions had to be measured relative to one another rather than to any external framework 'at rest'. He designed a new universe in which all observers, no matter how fast they were moving relative to each other, would always perceive the same universe in the same way, with the same speed of light. Absolute space and absolute time were replaced by the hybrid concept, space-time, and the magic *c* joined Newton's G (the Gravitational constant) and Planck's *h* as a basic property of the universe.

Before Einstein, *c*, like any other velocity, had been regarded as a function of space and time. It was a measure of how many metres of absolute space were covered per second of absolute time. Einstein reversed this, so that length and duration were not the determinants of velocity but the products of it. The length of a metre thus became dependent on where it was measured from. Similarly, the length of a second became dependent on where it was measured from. Two identical spaceships moving apart from one another at near-light-speed would measure each other differently – each would measure that the other had become shorter, and each would measure that aboard the other ship time was passing more slowly.

This notion defies common sense, but it is quite convenient for Einstein's conviction that an observer travelling close to the speed of light relative to some other point should see the same kind of universe as an observer at that point. The faster an observer goes trying to catch up with light-waves from his point of origin the shorter he gets and the slower time becomes (relative to that point of origin), so that the light-wave always appears to him to be covering the same distance in the same time. It is always beating him, literally out of sight.

It was not only time and space that Einstein dispensed with. Mass, as Newton had envisaged it, had to go too. Newton had conceived of gravity as a force by which masses attracted one another: he envisaged the planets orbiting the sun as if they were connected to it by a taut string, with the energy of their motion tending to pull them apart while the force of tension in the string held them together. After Maxwell, however, such illustrative concepts of force were dispensed with, and forces became fields of force – an essentially geometrical concept. (According to Mach, the concept of force, like space and time, was meaningless.) It was in making the force of gravity into a mathematical function that Einstein recruited Riemann's geometry. He conceived of gravity as a kind of curve in space – and in order to accommodate such a curve he had to add a new dimension to the universe. Mass – which is to say, the property of gravitational attraction – thus ceased to be an absolute and became a variable just like length and duration. Thus, the two identical spaceships moving apart at near-light-speed would not only measure a discrepancy in their lengths and durations, but also in their masses. While they watched one another's length contracting and time getting slower they would perceive one another's mass increasing.

It is vital to realise that all velocities in Einstein's system are *relative*. One often sees statements to the effect that a spaceship cannot go faster than light because its mass becomes infinite. This is the wrong way to look at it. The reason the spaceship cannot go faster than light is because no matter *how fast* it is going light is always going 3×10^8 metres per second faster. The ship can accelerate relative to its point of origin but *it cannot accelerate relative to the speed of light*. The time and the length and the mass that men on the spaceship measure does not alter – it is the time, length and mass which the people back home are *attributing* to them by measurement that is different. From their point of view the ship can go almost as fast as light, but it can never actually reach light-speed because they would only be able to perceive it approaching c – and they would be able to watch it approach c forever, because from their point of view ship-time would be slowing down to a standstill.

From the ship's point of view there would be no progress whatsoever towards the speed of light: as measured by the ship c remains the same (the ship is far worse off than the Red Queen in *Through the Looking Glass*, who only had to run twice as fast to get somewhere).

Einstein's theory of relativity made nonsense out of virtually all the concepts of classical physics. As an idea, it was most spectacular, but it was, at first, no more than an idea. Why should anyone have accepted it as an authentic account of the way the universe worked?

The main reasons why the theory was accepted initially, as a reasonable proposition, were reasons of logical aesthetics. As Occam's razor demanded, it did away with a whole host of imaginative inventions necessary to common-sense explanation. In particular, it dispensed entirely with the ether, and put the Michelson-Morley experiment in a new light.

That, of course, was not enough. The philosophy of science

demanded that Einstein's theory prove itself by making correct predictions about measurable phenomena. And it did. The points at which relativity can be checked against reality are relatively few, and they are rather abstruse points – but they do exist, and relativity *has* checked out. Einstein's theory accounted for certain disturbances in the orbit of Mercury, and it predicted that light-rays affected by gravity would displace the apparent position of stars almost in line with the sun (a claim which Eddington first substantiated during the eclipse of 1919). The gain in mass of highly accelerated particles produced by cyclotrons is appreciable enough to be measured. Einstein's theory has been confirmed by observation – but this should not be allowed to obscure the fact that the discovery itself arose by an aesthetic process from a fashionable way of thinking.

The revolution in science which Maxwell began and Einstein carried through to its logical conclusion consisted of the realisation and the revelation of the limitations of our powers of observation.

Science, in its remote beginnings, adopted as its mission the rationalisation of the world which we experience through the medium of our senses: the early Greek theorists sought explanations of the behaviour of the sun in the sky and the fact that the stars shone at night. And science maintained this purpose up to and including Maxwell's time, the late nineteenth century, when – at last – the final breakthrough seemed to be taking place. The scientists were concerned with explaining why the world that we can see is the way that we see it. Where they dealt with things *not* directly accessible to the senses, like magnetic or gravitational attraction, they were ready to assume that they could be interpreted in the same way – that is to say, that they were exactly the same as things we can see, but merely happened to be inaccessible to our senses.

Thus, atoms were at first visualised as tiny billiard balls, and later as miniature solar systems; forces were like tentacles or taut strings; light-waves were like ripples in a pond.

But Maxwell, and those who followed him, proved that this was not so. The things inaccessible to our senses did not, for the most part, behave like the things that we can see. To imagine, as many nineteenth-century scientists did, that the whole universe was merely an extension of the world before our eyes, and that once we could interpret what we could see then the riddle of the universe was answered, was gross arrogance. The universe is not just an infinite extension of the world that our senses experience.

Beyond the spectrum of visible light is a vast spectrum of electromagnetic energies. Within the margins of the smallest distance we can perceive – with the naked eye or with the most powerful microscope – there is a microcosm of great complexity. The world we see is both incomplete and superficial – and it is quite unreasonable to suppose that we can understand the whole universe simply by likening it to the tiny fragment of existence where we live.

There is a certain emotional reaction against the idea that we must abandon the obvious ideas which serve so well in making sense of our own world simply because they cannot be extended into the realms of the imperceptible. One can almost imagine people primly stating that if the universe cannot behave sensibly then they see no reason to associate with it socially – and perhaps this is one of the reasons for the retreat from modern science. Part of the demand for the popularisation of science is undoubtedly a demand for it all to be made easy – for complex ideas to be made accessible by analogy. Where analogies can help, they are very useful, and I am all in favour of attempts to make these ideas more accessible, but in doing so we must not overlook the basic point – that the necessary shift in perspective is not easy, and cannot be served by analogies alone. It is not easy to identify and recognise the illusions to which subjectivity makes us prey, for this is a purely intellectual effort – no matter how much we learn, our senses will still tell the same story.

Before Maxwell the truism that 'seeing is believing' was the anchor of science. But Maxwell introduced a new referee: the logic of mathematics. Mathematics, as a purely abstract discipline, is not subject to the same limitations as the human imagination. We cannot visualise a four-dimensional object, but mathematics can describe it. We cannot conjure up a picture of a photon, but mathematically we can define one. The mathematics we need in order to do these things is complex and abstruse, but are we to turn our backs on the universe and refuse to acknowledge truth simply because the square root of two is not a vulgar fraction?

Twentieth-century science has become infected with a new aesthetics – the aesthetics of mathematics. Einstein refused to tolerate 'ugly equations'. When Schrödinger first formulated his wave-equation he discarded it because it did not fit the known facts – and was later criticised by Dirac on the grounds that he should have known that such a beautiful equation could not be wrong; there had to be something wrong with the data. Bertrand Russell and A. N. Whitehead, in their *Principia Mathematica*, attempted to put all logic on a sound mathematical basis (and twenty years later Godel produced a mathematical proof of their failure to do so).

Elementary education has recently been invaded by the basics of a 'new maths', but for most of us mathematical aesthetics remains *avant garde* (we know nothing about maths but we know what we like). I do not think that it is necessary to acquire such new aesthetic sensibilities before we can begin to feel at home in the world of modern science, but I do think that we should notice its existence and consider its implications. Only then will we be properly equipped to explore the surrealistic realms of the microcosm and the macrocosm.

In closing this chapter I should simply like to quote the following from Sir James Jeans' book *The Mysterious Universe*:

Many would hold that, from the broad philosophical standpoint, the outstanding achievement of twentieth-century physics is not the theory of relativity with its welding together of space and time, or the theory of quanta with its present apparent negation of the laws of causation, or the dissection of the atom with the resultant discovery that things are not as they seem; it is the general recognition that we are not yet in contact with reality. To speak in terms of Plato's well-known simile, we are still imprisoned in our cave, with our backs to the light, and can only watch the shadows on the wall.

2

The Atomic Wilderness

The world of the atom is a world we know only by inference. We cannot perceive events in the microcosm, but we can contrive effects by which to detect and study them. Though we cannot see the subatomic particles themselves we can see the tracks which they leave as they pass through cloud-chambers: a wake of condensation which precipitates out around the ions formed as electrons are stripped from atoms by the speeding particles. But these contrived effects are all that we can see and therefore all that we can measure. Everything we know about subatomic particles has to be deduced from their interactions with grosser pieces of matter. This is no mean task, and it puts severe limitations on the kind of thing we *can* know about microcosmic events. It is rather as if an alien intelligence were to try to comprehend the human world on the basis of aerial photographs taken from forty thousand feet and nothing else.

In the last chapter I tried to illuminate the difficulties involved in thinking of light as composed of particles or waves or some combination of the two. Einstein's photon is none of these things – it is an assembly of mathematical properties, without substance: a mere ghost. Einstein's arrows, like Cupid's, are quite metaphorical.

The same thing applies to the subatomic particles. Because we term them particles, and because they possess mass, we tend to think of them as being rather more solid than electromagnetic energy. They are not. This does not mean to imply that there is anything unreal about them, merely to say that the criteria which we tend to use to define real things in the sensory world do not apply to them – I have termed them surreal, but those who prefer to invent a new word rather than adapt an old one might prefer 'altereal'. Either way, it is important to realise that atomic particles are not tiny lumps of primal matter but organisations of various behavioural properties. At this level, mass is simply one such property – a mathematical attribute. Solidity is a property of matter *in aggregation*, not of mass *per se*, just as the picture reproduced in a newspaper photograph is a property of an assembly of dots, not of the dots themselves.

It is difficult to get rid of the notion that because the proton has a mass 1,836 times that of the electron it bears pretty much the same relation to it as a bag of sugar does to an aspirin tablet. But the proton and the electron are ideas rather than objects, and it is worth remembering that although scientists have only recently realised that the 'thingness' of the world is superficial, and that reality is made of ideas, many poets (and even the odd philosopher or two) have always known it.

The first subatomic particle to be discovered was the electron. It was first introduced as a hypothetical concept by analogy with Dalton's atomic theory of matter: it was named by Stoney in 1891 as the 'atom' of electricity, but – as I have already commented – it remained a controversial concept for some years. Long before then, however, Plücker had called attention to a 'beautiful and mysterious green glow' produced by electrical discharges in evacuated tubes. It was concluded that this glow was the result of rays emanating from the negative terminal, which were therefore termed cathode-rays. In 1871 Varley suggested that these rays consisted of 'attenuated particles of matter' but it was not until 1897 that Sir Joseph Thomson measured the mass of the particle; he found it was 1,836 times smaller than that of the hydrogen ion.

The association of the electron with the atom was made in the early years of the twentieth century. Rutherford, studying the radioactive decay of uranium, identified different types of radiation with different penetrative power. One type – the alpha ray – was eventually identified as consisting of doubly ionized helium atoms, while another – the more penetrative beta-ray – was shown to consist of electrons. Thus it became known that atoms contained electrons.

The atom was first envisaged as a positively charged sphere containing 'loose' electrons, but this model proved incompetent and was eventually replaced by the Rutherford-Bohr model, in which the hydrogen atom was seen to consist of a single nuclear particle – the positively charged proton – around which an electron revolved in one of several possible orbits. More complicated nuclei thus became greater assemblies of protons, with more electrons orbiting them, and ions were formed whenever outer electrons were lost through one of a variety of subatomic incidents.

The important feature of Bohr's atom was the concept of orbits. By means of these orbits Bohr sought to explain the structure of the hydrogen emission spectrum: the characteristic pattern of light-frequencies emitted by incandescent hydrogen gas. This pattern was not continuous, like the colour-spectrum derived from the refraction of white light, but consisted of a series of lines, each line corresponding to a particular frequency. The distribution of these lines had been known for many years to follow a series generated by dividing a particular frequency by 1, 4, 9, 16, etc. (i.e. the sequence of the squares of the natural numbers) and commonly known as 'Balmer's Ladder'.

Bohr suggested that the reason the hydrogen atom could only radiate light at these particular frequencies was that these were the precise frequencies whose quanta were exactly sufficient to move the electron from one orbit to another, and that the electron could not exist between orbits.

This piece of rationalisation was extremely important. It had been known for a long time that each element in the periodic table had a characteristic spectroscopic 'signature' – by this means spectroscopic analysis had determined the composition of stars – but with Bohr's theory this signature assumed an entirely new role in atomic physics. Atomic physics after Bohr became a sort of spectroscopic graphology: the analysis of the hydrogen signature became the primary source of information about the world of the atom. Once again, it seemed that the key to the harmony of nature was to be found in whole numbers, for – thanks to Balmer's Ladder – Bohr had to introduce the concept of 'quantum number' to describe the possible orbits which the electron might assume.

There was, however, more to the hydrogen spectrum than first met the eye. As far back as 1896 Zeeman had discovered that the spectrum emitted by hydrogen atoms glowing in a powerful magnetic field became more complex – the simple lines split up into three, and sometimes more. A German physicist, Stark, found that electrical stimulation could result in individual lines becoming thirty-two or more.

In order to deal with these complications more factors had to be introduced into the theory. One set of whole numbers – those from which the Balmer sequence was derived – was not enough. There had to be others, and by 1916 the formula for calculating possible electron orbits contained not one quantum number but three. The number of possibilities open to the electron were increasing rapidly – far too rapidly, for as well as predicting all the actual spectral lines, the modified Bohr formulae also predicted a lot that were not there. In order to abort all the extra lines, Bohr had to introduce another modification into his theory: the so-called correspondence principle.

All this work was pure fudging. It is rather like the situation where a school textbook has the answers to all its problem questions in the back, so that a schoolboy in difficulties with set work is not actually faced with trying to find the answer, but with puzzling out how to get to the answer from the problem. Bohr knew the answer he was trying to find: the hydrogen spectrum, and he had the 'given' elements of the problem: his own atomic model. He simply continued adding new mathematical principles governing the behaviour of the electron in order that it should do exactly what it was supposed to do, and nothing else. But, try as he might, he never quite managed to get a 'perfect' equation which would describe exactly what the electron did do.

A fourth quantum number was introduced, and then Goudsmit and Uhlenbeck suggested that much that was odd could be explained by assuming that the electron had a spin of its own as well as an orbit round the nucleus. By the mid-1920s the situation was both rather desperate and rather comical. Physicists everywhere were playing the game of 'hunt the equation . It was one of the most important searches in the history of science, and it was carried out entirely in the imagination of the men involved. It is perhaps not entirely surprising that four men came up simultaneously with four different versions of the answer.

The first of the four was de Broglie. His contribution was conceptual, for it was he who stated that physicists must stop viewing the electron purely and simply as a particle. He said that the electron had wave-properties as well as particle-properties and – like the photon – must be regarded as a purely mathematical entity. At the same time, purely in the interests of fudging the figures, Heisenberg, Schrödinger and Dirac were all applying the mathematics of wave behaviour to the problem of producing a better equation to capture the essence of electron behaviour. They all succeeded in producing a new mathematical programme for the description of the atom – each of them using different mathematical tools. Their separate work was done during 1925 and 1926, but by 1928 Dirac had absorbed the other theories into his own – had shown that all the statements contained the same truth, and were variants of one another. The coincidence was so striking, and the harmony of the mathematical models so wonderful, that the rightness of the new theory was beyond aesthetic doubt. The only trouble was that it went much further beyond the limits of common sense than anyone had gone before, for the new equations introduced not merely an unspeakable number but an imaginary one: the square root of minus one. Out with the common-sense notion of the particle electron went the common-sense notion of the electronic orbit. The orbit, like the particle, became a mathematical abstraction. The position and momentum of particles could no longer be calculated as single quantities, but as sets of properties, and there was no way to define them accurately – indeed, they varied according to the order in which they were calculated, and it was in measuring that variation that the square root of minus one had to be invoked.

All this was to rationalise the hydrogen atom: the simplest atom of them all, consisting of just one electron and one proton.

At first, it had been thought that atoms contained only protons and electrons, and thus that all elements were compounds of hydrogen nuclei. The obvious flaw in this theory was that the atomic weights of the elements did not match the sum of the proton masses. Thus Rutherford suggested, in 1920, that there was another nuclear particle, neutral in terms of electric charge, which was present in all nuclei except hydrogen. He named this hypothetical particle the neutron, and supposed that its mass was approximately equal to the mass of a proton. Thus the helium nucleus (the alpha-particle), which contained only two protons but was about four times as heavy as a hydrogen nucleus (i.e. a proton), was tentatively identified as consisting of two protons and two neutrons.

The neutron was finally identified as a separate entity in 1930–2. By this time bombarding nuclei of heavier elements with alpha-particles to see what might be chipped off was a popular form of atomic exploration, and the neutron turned up as an extremely penetrative 'ray' emitted when beryllium nuclei bombarded with helium nuclei combined to form carbon nuclei, losing a neutron each to balance the nuclear equation.

In this period, as the case of the neutron demonstrates, experimental discoveries were lagging a long way behind theoretical ones. The neutron was invented as a theoretical convenience ten years before any hard evidence for its existence turned up. This pattern of events was to become typical – a theoretical physicist would explain away an imbalance in his equations by hypothesising a new subatomic particle whose properties conveniently arranged for the sum to come out right, and at a later date someone would find and identify exactly such a particle.

Perhaps the most remarkable instance of this method happened in 1931, in parallel with the discovery of the neutron. Pauli was attempting to balance the mass/energy equations involved with radioactive decay, and kept finding a 'mass deficit'. A uranium atom, for instance, breaks down to a thorium atom, emitting an alpha-particle. The mass of the resultants has to be less than the mass of the original anyhow, because some of the mass is converted into the energy which shoots the spare particles out of the uranium atom. The amount of mass converted to supply this energy can be calculated by Einstein's equation $E = mc^2$.

But Pauli found that when beta-particles were emitted the energy they possessed was rather less than the energy which ought to have been produced by the mass deficit. Some of the mass which was disappearing simply did not manifest itself as the energy of the particles – not, at least, of the known particles. The trouble was that if there was another particle involved it was impossible to detect with the equipment in use at the time.

Pauli, however, was prepared to be adventurous. If an undetectable particle was required to balance the books, then an undetectable particle there had to be. He thus invented the neutrino, which had no electric charge and no mass.

The fact that this hypothesis was taken seriously is most eloquent testimony to the change which had taken place within the scientific establishment. Fifty years earlier – perhaps even twenty years earlier – a physicist who calmly proposed that a gap in abstruse theory would have to be filled by supposing a new entity with properties all but unimaginable would have seemed absurd. But in the new intellectual climate it seemed not at all out of order to suggest that if the observations of the experimenters did not provide perfect equations, then the observations had to be deficient, and deficient by precisely the amount required to satisfy mathematical aesthetics.

Pauli was taken seriously – so seriously that experimental physicists set out to comply with the demand and detect the fugitive particle. This was no easy matter. Like all atomic particles the neutrino could only be detected and identified by measuring its interaction with some grosser organisation of matter – but the probability of an interaction between a large nucleus and a massless, chargeless particle with sufficient energy to carry it clean through virtually any amount of solid matter was very small indeed. A very small probability, however, remains finite, and the experimenters were certain that if they looked long enough and hard enough they would succeed.

The first neutrino-detecting experiments attempted to catch neutrinos from an atomic pile in long tanks of water laced with a chemical whose molecules emitted sparks when struck by subatomic particles. The long tanks were surrounded by hundreds of phototubes sensitive to the smallest scintillation. The trouble with this experiment, however, was that other kinds of particles were being emitted and interacting as well. But when the number of interactions detected was added up and compared to the expected numbers based on theory there was sufficient correspondence to suggest that a few neutrinos had, in fact, been trapped.

In order to obtain unequivocal confirmation of the neutrino it was necessary to filter out all the other particles which were confusing the issue. In order to do this the water-tank apparatus was set up in mine-shafts, using more than a mile of the Earth's crust as a filter to trap all the less energetic particles. (Neutrinos are so unimpressed by solid matter that most of the ones emitted by the fusion reaction in the sun go clean through the whole Earth.) Over a period of years, the experimenters finally became convinced that they had captured a few neutrinos – six in all. Thus (at tremendous financial cost) was mathematical aesthetics vindicated.

There was, however, more to atomic physics than simply balancing equations. The equations provided an exact mathematical description of *what* was happening in the atomic world, but they could not explain *how*.

It was obvious that there were unknown forces at work in atomic nuclei – the fact that the nucleus stuck together at all was ample evidence of this. Protons in the nucleus repel each other with an electromagnetic force much more powerful than the gravitational force which tends to keep them together, and there had to be a much more powerful force – a 'strong nuclear interaction force' – overriding the electromagnetic repulsion.

(Gravitational force, which is so important in the macrocosm, only seems so powerful because gross aggregations of mass tend to be electrically neutral – the charges balance out – and because the nuclear forces only work at very close range. Gravity is, in fact, the weakest of the known forces.)

Inventing a new force is no more difficult than inventing a new particle – but it is no less difficult, and just as Pauli had had to provide a theoretical recipe for his hypothetical particle, so there had to be a rational description of the way the strong nuclear interaction force operated.

It was already known that energy was communicated in electromagnetic force-fields by means of photons. Aesthetic considerations had also prompted the suggestion that the communication of energy in gravitational force-fields might be concerned with hypothetical photon-analogues: 'gravitons'. (No one has yet detected a graviton, but many physicists are prepared to credit this to the deficiencies of experimental detection techniques.) On the same basis, it seemed likely that the transmission of energy in strong nuclear interactions might require a photon-analogue, and to fill this role Yukawa, in 1935, designed the *meson*. He proposed that the nuclei of all elements more complex than hydrogen were held together by means of a constant exchange of energy between protons and neutrons. In this way neutrons, hitherto merely reservoirs of spare mass, were given a role to play in the atomic scheme of things. Yukawa calculated that the energy-exchanger ought to have a fairly considerable mass – about 270 times the mass of an electron – and thus ought to be conveniently detectable. (It was because of this mass – intermediate between that of an electron and that of a proton – that the particle was dubbed the meson.)

A few years later, as per prescription, a meson was detected. Carl Anderson found in some photographs of cosmic radiation traces in a cloud-chamber an anomalous track. The kind of trace which the particle had left allowed its mass to be calculated, and it was, indeed, intermediate between the mass of an electron and the mass of a proton, although somewhat lower than Yukawa's predicted figure. The discrepancy was shunted aside – the power of the prediction/discovery pattern was by now so strong that no one doubted it had been fulfilled yet again.

Perhaps the most remarkable of all the hypothetical entities later discovered in actuality was the positron.

This particle had been 'invented' by Paul Dirac, the mathematician who had wrapped up the theory of the hydrogen atom in the 1920s, integrating his own work with that of de Broglie, Schrödinger and Heisenberg to make a marvellous new synthesis. Dirac was a great believer in the mathematical perfection of the microcosm (I have already referred to his criticism of Schrödinger, who doubted his equation rather than the known facts when the two seemed at odds) and he was possessed of the most highly developed sense of mathematical aesthetics.

He invented the positron not because there was any gap in theory which called for the existence of such a particle, but simply because his equations *allowed* the existence of such a particle. The positron was simply an electron with a positive charge: a 'counter-electron' or 'anti-electron'. He suggested, in fact, that there might be a whole spectrum of anti-particles, each one the electromagnetic counterpart of a known particle. It was not theoretically necessary that such particles should exist, but it was mathematically plausible: his equations worked just as well with all the signs reversed.

In a sense, the idea of anti-particles was a purely intuitive one – an inspired guess. But it is significant that it was Dirac who suggested it, and there is really more to it than intuition: it reflects the deep faith which Dirac had in the genuineness of the mathematical principles underlying atomic physics; a confidence that *the universe would conform to the maths* rather than the maths simply fitting the universe. Dirac had faith in the ultimate harmony of nature.

And Dirac was justified. He predicted the existence of anti-particles in 1930, and in 1932 Carl Anderson (discoverer of the meson) identified positively charged electrons in the cloud-chamber tracks he was studying.

Dirac had also predicted that anti-particles would have a very short lifetime in matter, as they would meet ordinary electrons, resulting in mutual annihilation. This prediction, too, proved to be true.

The logic of symmetry suggested that there ought to be an equal number of anti-particles and particles in the universe (as, mathematically, the two are equally likely). The relative scarcity of anti-particles locally was easily explained by Dirac's second prediction – the only reason that matter can exist at all is that there is so very little 'anti-matter' around. Photons, however, are ubiquitous, and would exist just as well in anti-matter systems as they do in matter systems, and it is theoretically possible that other stars or other galaxies might be composed of anti-matter rather than matter, and that therefore there might be an equal number of particles and anti-particles in the universe as a whole. This remains pure conjecture, but might be an attractive idea to those with Dirac's faith in the mathematical balance of nature.

At the end of the 1930s the theoretical physicists seemed to have triumphed. Their attitude and their approach seemed to have been vindicated by experimental results. The work done in the laboratory had so far been concerned almost exclusively with testing the predictions of the theorists – much as Hertz's research into radio-waves had been undertaken simply to prove Maxwell right, and not with any practical end in mind. But the coming of the Second World War was to put atomic research into a new light, and the practical possibilities innate in atomic theory were revealed in the most spectacular way: they killed a hundred thousand people at a single blow.

The principle on which the atom bomb was based was consummately simple. Uranium nuclei decay of their own accord, releasing energetic particles. If those particles collide with the nuclei of other atoms, those too may break up, releasing more energetic particles and so on, until millions of nuclei breaking down simultaneously release a vast flood of energy. Such a reaction can be started simply by bringing together a mass of pure uranium-235 so large that the probability of the energetic particles liberated by spontaneous decay causing further decay becomes very high. The energy liberated is then fed back into the reaction, causing it to go ever faster. The mass sufficiently great to permit this to happen is known as the critical mass.

From the point of view of the theoretical physicists an atomic chain reaction was simply another possibility permitted by their mathematical models. But in the real world, it meant something very different, and both the nature and the tempo of atomic research were changed because of it. Funds became liberally available for research in the laboratory.

The allocation of a great deal of money to atomic research, and the encouragement of governments, made little, if any, difference to the pure theorists like Dirac, who worked within their own imagination. But it made a big difference to the experimental side of the research. There was rapid development of equipment, and far more equipment in use.

The standard Wilson cloud-chamber was largely replaced by the bubble-chamber, in which the water-saturated gas was replaced by a gas-saturated liquid. The 'trails' formed in the same way – by precipitation around ions whose electrons had been stripped away by speeding particles – but the bubble-chamber was rather more efficient at trapping particles. At the same time, the techniques used to record and measure particle-tracks were refined greatly – photographic emulsions sensitive enough to record the tracks of even the most rapidly moving and evanescent particles were developed and applied.

It was this great boost to experimental technique that enabled the experimental physicists to catch the neutrino – and the new financial interest in atomic research which permitted them to do it. But the boost accomplished very much more than that, for it enabled the experimenters to far outstrip the theoreticians in making discoveries, and changed the whole complexion of twentieth-century physics. Within a decade from the end of the war, physicists had not only discovered most of the particles predicted by the imaginative theoreticians, but nearly a hundred new particles which, not having been predicted, seemed surplus to requirements.

This put the adventures of Pauli, Yukawa *et al.* in an entirely new light. In the 1920s and early 1930s it had seemed bold and ambitious to take care of discrepancies in theory by inventing new particles – it was contrary to the spirit of Occam's razor and dead against the philosophy of Mach: a possibly needless multiplication of hypothetical entities. But the business of designing one's hypotheses to fit the gaps in one's theory cuts both ways – the risk that one will come up with one particle too many exists alongside the risk that the action of a whole host of particles will be lumped together as one.

Post-war experimenters found the particles that pre-war theorists had imagined, but at the same time they smashed the assumptions which lay behind the attitudes encouraging the theorists to make the predictions in the first place. The basic faith of the aesthetic mathematicians in the economy of nature and the elegance of the fundamental organisation of the universe began to seem rather unjustified.

One of the first of the new particles to turn up (in 1947) was, ironically enough, the Yukawa particle. The discrepancy in mass between the meson Anderson had discovered and the one Yukawa had predicted had been dismissed at the time as unimportant, but now it became obvious that there was more than one kind of particle intermediate in mass between electrons and protons: a whole family of mesons. Anderson's particles were named mu-mesons (muons) and the new mesons pi-mesons (pions). It appeared that pions decayed very quickly into muons (losing a neutrino in the process) and that muons were also extremely short-lived, decaying into an electron and two neutrinos. Another meson − the K-meson − discovered at about the same time (more massive than the pion) later turned out to belong to a third distinct family of particles.

By 1953 various groups of observers had detected particles more massive than protons, which − like the mesons − decayed very rapidly into other high-velocity charged particles. It was, by this time, necessary to discover a convenient categorisation of the ever-growing collection, and they were provisionally split into four classes depending on mass. The lightest particles, including the electron, were classed as *leptons*, the middle-range particles kept their designation as *mesons*, particles in the proton-neutron mass range became known as *baryons* and the heaviest particles were dubbed *hyperons*.

By the late 1950s the priority had shifted from particle-hunting to the attempt to discover in the vast array of named particles some semblance of order: a kind of 'periodic table' of atomic constituents. In 1961 Gell-Mann and Ne'eman suggested that some hyperons formed a kind of hierarchical family. They identified four delta-particles whose masses were approximately 2,200 times the mass of an electron, with charges − 1, 0, + 1 and + 2; three sigma-particles with masses just over $2,300m_e$ and charges − 1, 0 and + 1; and two xi-particles with masses approximately $2,600m_e$ and charges − 1 and 0. They suggested that to complete the family there ought to be one more particle, with an even greater mass and a charge of − 1. After an intensive search, the omega-particle was discovered to meet these requirements − the first triumph for aesthetic prediction in some years.

As yet, no one has emerged who can do for the subatomic world what Mendeleev did for the chemical elements and offer a *total* pattern of organisation for the elementary particles. Perhaps it is unreasonable to expect that one may be produced − after all, but for a couple of eccentricities Mendeleev's pattern is a simple two-dimensional array: a product of plane geometry. We know that such approximations are practical and useful in our immediate cosmos, which is one layer of the microcosmic/ macrocosmic spectrum, but we know that applied to the cosmos as a whole plane geometry is very limited in its applicability.

Gell-Mann and Ne'eman's classification of the ten hyperons is, like the periodic table, a two-dimensional array, with mass and charge the two factors varying. But mass and charge are only two of a number of properties which can be used in distinguishing elementary particles. There is *isospin* — another property, apart from charge by which the proton differs from the neutron. There is also *strangeness* — a property attributable to particles involved in the 'weak nuclear reactions' (which also involve a fourth class of force, more powerful than gravity but less powerful than electromagnetic or strong nuclear forces: strangeness is the property through which weak nuclear force is expressed, just as gravity acts through mass and electromagnetism through charge). Both these properties can be expressed by a simple array of whole numbers (like charge) but it is not clear how far the arrays might extend. Particles with strangeness numbers of 1, 0, −1 and −2 are known, but isospin seems to be an either/or phenomenon.

In the recent past new particle-identifications have led to the suggestion that even the list of properties is not yet complete. A particle discovered independently by two groups in America, and named variously the psi-particle and the J-particle, may be possessed of an entirely new property. It has been suggested that the new particle be called 'charm' because the particle seems to have an unusually long lifetime (a 'charmed life').

All this is remarkably complex and bewildering, and it represents a wholly unwelcome complication of a situation which once seemed so simple and so mathematically elegant. The attempt to rationalise and systematise these discoveries, however, goes valiantly on. It is hoped that the four different kinds of force-field we now know are, in fact, the whole set. It is hoped that all four operate on a basically similar pattern, each involving an 'exchange particle' (though the gravitational exchange particle — the graviton — and the weak nuclear force exchange particle — the W-boson — remain hypothetical). One cannot help feeling, however, that just as the physics of the early part of the century exposed the limits of the old logic, the physics of the post-war era is testing the limits of the new.

Virtually all the subatomic particles have very short lives. The xi-zero-particle, which lasts for 1/10,000,000,000 of a second, is an unusually stable particle − the sigma-zero-particle, which is the least stable of the hyperons, has a lifetime approximately a million times less than that.

It is not easy to imagine time-spans so short, or to see how the fabric of the universe is maintained by complex events taking place on such a time-scale. We must, however, remember that time and space are different aspects of the same thing, and that time-scale is related directly to size-scale. The linking factor is, of course, the velocity of light.

SIZE in cm (order of magnitude)				TIME in secs (order of magnitude)			
−14		9	Earth	−25		−2	
−13	Proton	10		−24		−1	
−12		11		−23		0	
−11		12		−22	Omega(−)	1	
−10		13	Earth's orbit	−21	meson	2	
−9		14		−20		3	Neutron
−8	Atom	15	Solar system	−19		4	
−7		16		−18		5	
−6	Small virus	17		−17		6	
−5		18		−16	Sigma$_o$	7	Calendar year
−4	Bacterium	19		−15		8	
−3		20		−14		9	Human
−2	Amoeba	21		−13		10	life-span
−1		22	Galaxy	−12		11	History
0	Acorn	23		−11		12	
1		24		−10	Xi-zero	13	
2	Man	25		−9		14	
3		26		−8	Pion	15	
4		27		−7		16	
5		28		−6	Muon	17	
6	Mount	29	Visible	−5		18	Star's
7	Everest	30	universe	−4		19	life-span
8		31		−3		20	Probable life
9	Earth	32		−2		21	of universe

Figure 2   Size- and Time-Scales in the Microcosm and the Macrocosm

Figure 2 shows a comparison of size-scale and time-scale in the microcosm with size-scale and time-scale in the macrocosm, the parities in each case being defined by the velocity of light. The analogy between the two systems is drawn by comparing the size of an atom to the size of the solar system. (One has to be careful of such analogies lest they foster the old illusion that an atom is like a solar system, with electron planets circling a nuclear sun — I must make it clear that it is only the *size*, and not the *nature* of the systems that I am comparing. As in Figure 1 the scales are measured in orders of magnitude, so that both size-scales and both time-scales are, in fact, parts of the same infinite spectrum.)

It will be seen from the diagram that though the vast majority of subatomic particles seem to us to drift in and out of existence in no time at all, the situation is very different when seen in context. Even the evanescent sigma-zero-particle lasts as long as it takes light to travel three hundred atomic diameters.

We have now drawn a comparison of the scales of microcosm and macrocosm, and pointed out that the two are simply different ranges of a single spectrum. We have already pointed out, in beginning this programme of exploration, that the universe we perceive is only a thin slice of reality. It therefore seems reasonable to suggest that even the microcosm and the macrocosm — the 'reality sandwich' we are discovering by inference — are only a slightly thicker slice.

This is not a new idea. When the atom/solar system analogy was initially drawn, there was rapid speculation (on the part of writers of scientific romance if no one else) as to whether our solar system might not be merely an atom in a greater universe, and that atoms might be tiny solar systems, whose atoms were tiny solar systems, and so *ad infinitum*. Ray Cummings, one of the early writers who happened upon this idea, claimed to have been inspired by an advertisement for Quaker Oats, which showed a quaker holding the packet, on which was a picture of the quaker holding the packet, etc., etc.

We now know that such a simple *structural* analogy is quite incompetent. But the *scalar* analogy remains. And the question remains: are the 'elementary particles' really elementary? Or are there sub-subatomic entities and events?

It was in pursuit of this idea that while some theoreticians were searching for a periodic table of atomic constituents, others were searching for a new set of ultimate particles. The attempt to rationalise all atoms in terms of three constituents (the proton, the electron and the neutron) had failed, but it had come very close to providing a complete account of the periodic table. Was it not possible that all supposedly elementary particles might be rationalised on a similar system? Here will be recognised the same fond hope (or dogged faith) in the essential economy and harmony of nature we have met so many times before, and will meet again. In this case, the idea led to the invention of *quarks*.

The name 'quark' derives from a phrase in *Finnegans Wake* by James Joyce – a wry acknowledgment of the surreal nature of the exercise. Apparently, three quarks (and their inevitable counterparts demanded by the principle of symmetry, the anti-quarks) are enough to account for the differences between all the strongly interacting particles (the pion, the kaon, the proton, the neutron, etc.) though they seem to meet their match in the photon and the electron. Quark-hunters initiated the inevitable search, though – like the neutrino – a quark would be almost impossible to detect and identify, but they have met with little success.

The most interesting thing about quarks is not their peculiarities (they are for instance, reputed to have fractional charge and their individual masses are considerably greater than the masses of the particles they combine to make) but the fact that scientists should have felt it necessary to invent them at all. The quark hypothesis is not only a rather desperate attempt to find some sense in it all, but also a perfect example of the kind of sense which scientists now expect to find in it all – the kind of truth they now believe in. It is a fugitive truth – a truth with its own in-built irony.

The quark hypothesis represents the limit of our scientific aspirations. We can go no further, not because there is no further to go but because we have no means of going. The understanding which we have of the way the world works can be extended beyond the world we perceive into worlds which we know by inference. But there is a limit to what we can learn. We can only go as far into the microcosm and the macrocosm as we are permitted by the yardstick which we use for measuring events. That yardstick is *c*, the velocity of light: it is not only the *standard* by which we measure, but also the *means* of measurement. The photon is the ultimate constituent of our conceptual unit, for we have no information which is not communicated to us by photons (except in the gross world of our five senses – and even there sight is predominant).

It is for this reason that our knowledge of the microcosm is clouded with uncertainty.

I have delayed discussing Heisenberg's uncertainty principle until this point, although chronologically it belongs to the early part of the chapter, because I think it is easier to see its significance in the light of these later discoveries.

There are a number of ways in which the uncertainty principle can be expressed, but the one which best conveys its meaning and puts it into its proper context is the statement that *the act of observation alters the properties of that which is being observed.*

We can see how this principle is derived from the argument set out above by considering the thought experiment by which Heisenberg himself illustrated the principle. He imagined a single electron travelling in a vacuum past an observation point. The object of the observer is to determine its position and its velocity at a particular point in time. But the only way the experimenter can make any observation at all is to 'see' the electron: which means, in effect, bouncing a photon off it. The trouble is that bouncing a photon off an electron is certain to interfere in some way with the quantities which the observer is trying to measure. In order to determine its position *exactly* the observer has to use a photon of sufficient energy to knock the electron off course once the quantity is determined – thus, once the position is measured the velocity no longer can be. On the other hand, to measure the electron's velocity he has to use photons of very low energy to track its progress, so as not to deflect it – but then measurements of its position in space become equivocal.

Analogues of Heisenberg's uncertainty principle in the everyday world are not hard to find. Psychological and sociological experiments of all kinds tend to be victims of the principle – however the experimenter tries to collect his data he will probably interfere with the processes he is trying to study. Experimental biology also provides obvious instances – like the man who goes out to study the habits of nocturnal animals armed with a powerful torch so that he can see what they are doing. There is also the famous anecdote about the man who left a monkey alone in a room full of apparatus, and then peeped through the keyhole to see what the animal would do. The monkey came to the door and peeped through the keyhole to see what the experimenter was doing.

There is an ancient philosophical conundrum which asks whether a tree falling in a great forest where no men live makes any sound as it falls. In other words, is a sound which no one hears still a sound? Are the things around us still there when we are not looking at them? Common sense answers 'yes'. When the tree falls the air vibrates even though there is no ear to sense the vibration. The things around us are, we believe, real, and their existence is not dependent upon there being someone to perceive them. Common sense, therefore, suggests that Heisenberg's uncertainty is subjective. We cannot measure the position and the velocity of an electron, but if the observer were to let the electron alone and refrain from bouncing photons off it, it would still *have* a position and a velocity, wouldn't it?

Well, in a word, 'No'.

We have already seen that the logic of common sense *cannot* be extended to the microcosm. The electron cannot be said to have a position and a velocity of its own – only a position and velocity relative to the observer. Heisenberg's thought experiment is very much akin to Einstein's: just as the spaceship travelling at near-light-velocity relative to the observer on Earth has, *relative to* the observer, become foreshortened and more massive, and appears to have changed gear in time, so the electron has become uncertain in its position and velocity. The uncertainty is relative, but real. We cannot extend our assumptions about 'where' and 'how fast' to the electron any more than we can extend our assumptions about 'where' and 'how fast' to the spaceship.

Heisenberg discovered that the Bohr model of the atom could never provide a full explanation of the spectroscopic signature simply because the nature of the explanation which Bohr was seeking was inapplicable. Bohr wanted to explain the spectrum on the basis of where the electron was and where it might go. Heisenberg discovered that in actuality there was no 'where' – only a series of 'might be' wheres. This dovetailed exactly with de Broglie's suggestion that the electron might behave like a wave rather than a particle, and Schrödinger's wave-quotation, which provided a mathematical description of that facet of its behaviour.

But if where means so little with reference to the microcosm, how is it that it has such a firm and practical meaning in the intermediate cosmos – the perceived world, which is, after all, made up of atoms? Basically, it is because of statistical aggregation. Just as Maxwell discovered that the temperature of a gas is determined by the average kinetic energies of its molecules, so Heisenberg discovered that the properties of gross matter are determined by the average 'whereness' of its constituents. Just as temperature can be measured with great precision with respect to a gas, but becomes uncertain when applied to individual molecules, so the behaviour of large numbers of electrons can be defined although that of individual ones cannot.

There is, therefore, a basic uncertainty not only in our crude perception of the universe, but in the universe as it exists relative to us. Mathematically, as Heisenberg and Dirac found, the uncertainty can be expressed by a function involving the square root of minus one. There is, I think, a certain propriety in the fact that Heisenberg conducted an imaginary experiment and arrived, eventually, at an imaginary result. But imaginary, in this case, does not mean unreal.

There now remains only one major discovery made by the modern explorers in the microcosm which needs comment here. It is, in some ways, the strangest and perhaps, thereby, the most significant discovery of them all.

I have already commented on the important role played by the principle of symmetry in modern physical theorising. It seems such an integral part of nature, and − as demonstrated by Dirac and the quark hypothesis − it is intimately interwoven with our expectations about the form of ultimate truth. The mathematical plus and minus are mirror-images, and we expect, when we look into a mirror, to see everything in our own world reflected there − with left and right reversed, but exactly similar in all other respects. It seems reasonable to us that if the whole universe were, in fact the other way round it would really be just the same. Perhaps, however, we should remember that when Lewis Carroll sent Alice *Through the Looking Glass* the world she found there was very different.

In the human world, left and right are not at all equal. The human body, though superficially symmetrical, definitely has two distinct sides. In fact, this asymmetry extends throughout the world of living things: not only is there no perfectly symmetrical living organism, but the very chemicals of life are asymmetric. Many organic compounds exist in more than one form, depending on the grouping of the atoms in the molecule − these forms are known as *optical isomers* because they are molecular mirror-images: chemically identical, but different in the direction in which they rotate light rays shone through them. Chemistry draws no distinction between optical isomers, but biology does − virtually all the organic compounds in living tissue are laevo-rotatory (i.e. left-handed). To the chemist − and even more so to the aesthetic mathematicians and physicists − this inherent bias of living organisms may seem to be idiosyncratic, a flaw in nature which should exist only as an accident of gross organisation. The chemist does not expect to find such discrimination in the more rational world of basic chemistry, much less does the physicist expect it to extend into the objective world of mathematical understanding. Equations are not expected to be biased towards plus at the expense of minus, or vice versa.

In 1957, however, this attitude was challenged by two theoreticians – Yang and Lee. They suggested that there was a clear distinction between right-handed spins and left-handed spins in the subatomic world, and that one was favoured over the other. A remarkable experiment was devised to test this hypothesis, by which radioactive cobalt atoms were precisely aligned in a magnetic field, and the pattern of electron-emission as they decayed was observed. Instead of the electrons being impartially shot out both left and right, many more went one way than the other. It would not have been so bad if they had all gone the same way, but they did not – they *could* have gone either way, but more went left than right. Yang and Lee were right, and the universe appears to be left-handed.

This discovery is generally termed 'the fall of parity', and it appears to imply that the conservation laws – the physical equivalent of mathematical equations – do not always hold. The particular instance in which they do not, which was pointed out by Yang and Lee, is a very trivial one, and with a little juggling of mathematical concepts the failure of conservation can be covered up. The discovery has not exactly shaken the foundations of modern science, because the principle seems to hold good everywhere else, but what it does do is present a direct challenge to our assumptions about what *kind* of truth we might find at the end of our search for the answer to the riddle of the universe.

Mathematical aesthetics has served the twentieth-century physicists very well indeed – but Newton's mechanics served perfectly well for all practical purposes for two centuries, and still serves for most *practical* purposes. We have learned, however, that the assumptions which underlie it are wrong. The revelation of the fall of parity may suggest to us that perhaps the new physics is just as wrong, but merely suffices for a new set of practical purposes. It may be that we have been betrayed in our search for the harmony of the universe yet again – that the fall of parity is for modern science what the unspeakable square root of two was for Pythagorean doctrine.

Perhaps, in the ultimate analysis, such harmony as there is in nature is only approximate. Perhaps there is none at all, or if there is, perhaps it is such that we may never be able to come to terms with it because of the limitations of the photon as an information carrier. It may be that the *ultimate* truth is (to borrow a phrase from another context) in 'light inaccessible, hid from our eyes'.

3 The Egocentric Illusion

In this chapter I want to look at the ways in which modern science has affected our ideas concerning the intermediate cosmos – the fraction of the universe of which we have 'direct' sensory knowledge. Unlike the strange world of the atom and the further reaches of space, this is not a world which we have discovered recently, and attempts to explain it have been made since antiquity. In this area of thought ideas tend to have evolved slowly, and resistance to change has always meant that there is a legacy of old ideas mingled with the new. Despite the fact that in the exploration of the visible universe we have the evidence of our eyes to help us, we still have to rely to an inordinate extent on inference and imagination in gaining knowledge of it. In fact, the involvement of our senses often means not that we can be more sure of our evidence, but that we have the greatest of difficulty escaping our illusions.

There is one particular illusion which has influenced human attitudes to the world since time immemorial, and which continues to do so, although its presence within our assumptions is ever more covert, and that is the idea that we stand (literally, or in some way metaphorically) at the centre of the universe. In the pages which follow I should like to look at one of the ways in which this idea infects twentieth-century scientific belief – and how actual discoveries stand relative to its implications.

The first man to attempt to synthesise astronomy and physics into a rational account of how the universe is put together and the principles on which it operates was Aristotle. In Aristotle's model of the universe the human world was quite literally at the centre, with the heavens distributed in a series of spherical shells which revolved around it in a complex pattern determined by an intricate system of cosmic gears. The Earth, according to Aristotle, was the world: there was no other. While the Earth was made up of the four elements (earth, air, fire and water) of physical substance, the heavens were quite different – not material at all, but *quintessential* ('quintessence' means, in fact, 'fifth element').

Aristotle's scientific ideas are much maligned today by scientific historians, not so much because Aristotle was all wrong (anyone can be wrong) but because the Aristotelian model was adopted by the Christian Church as official dogma in the thirteenth century and was supported by religious faith in the intellectual battle against Copernicus's heliocentric system. Modern scientific historians often make something of a hero out of Aristotle's contemporary, Aristarchus of Samos, who suggested that the sun was the centre of the universe and that the Earth went round it, but Aristarchus's ideas were rejected then primarily because he could find no reasonable evidence to support them.

There is no need here to go into detail about the way in which Copernicus, Kepler and Newton fought for a heliocentric universe and won despite the staunch opposition of many of God's ambassadors on Earth. Suffice it to say that in the seventeenth century it was demonstrated conclusively (by arguments not available to Aristarchus of Samos) that the Earth was not, in fact, the centre of the universe.

As time went by, it was also demonstrated that the sun was not the centre of the universe either. Herschel and Kapteyn in the 1780s pictured a vast 'sidereal system': a lens-shaped agglomeration of stars of which the sun was but one (albeit pretty near the centre). By the early years of the twentieth century it was shown that there are far more stars in the lens than are visible to the naked eye, and that in fact the sun is nearer the rim than the centre, and at about the same time it became clear that the sidereal system was not a singular organisation of stars but one galaxy among millions.

This procession of discoveries led the physical centre of the universe progressively away from little old Earth, but it did not kill the egocentric illusion – merely changed its basis. Astronomical discoveries made it clear that the position of the Earth was not special, and spectroscopic analysis of the stars showed that it was not special in terms of composition, either, but was made of the same stuff as the rest of the universe. A new logic thus appeared to commend itself: the Earth was not different from the rest of the universe, therefore the rest of the universe must be much the same as the Earth.

One of the most unfortunate men caught up in the ideological war between church and science after Copernicus was Giordano Bruno, who was burned as a heretic in 1600. He had been inspired by the relegation of the Earth from its place of privilege to imagine an infinite universe filled with suns around which revolved worlds like the Earth, inhabited by men. He had leapt, imaginatively, from one extreme to the other: from the idea of a unique Earth to the idea of the Earth as a pattern for the entire universe. And it is *this* version of the egocentric illusion – that we can make assumptions about other worlds and other stars based on what we know of our own world and our own star – that haunts twentieth-century science.

By removing the Earth, conceptually, from the centre of creation to its status as an ordinary planet of an ordinary star of an ordinary galaxy, scientists tended to lose altogether any awareness of the Earth as a complex, strange, and possibly unique object: by becoming conceptually 'ordinary' it became conceptually simple. This is an illusion, and it is an illusion which the actual discoveries of twentieth-century science are having great difficulty in overcoming. We have been very slow to re-align our expectations to conform with our knowledge instead of our entrenched beliefs.

Confined as we are to the surface of the Earth – unable to dig more than a short distance beneath it or (until 1957) fly high above it – we have, in the past, been inclined to think that all the complexity of the Earth is contained at that surface, with only rock beneath and only air above.

It is, however, not so simple as that.

We are approximately four thousand miles from the centre of the Earth. Our deepest mines and wells go down a fraction over two miles. In our investigation of the Earth as an object, therefore, we have literally only scratched the surface. Virtually all we know about the solid component of the Earth we have deduced rather than observed.

The mass of the Earth was first determined in 1798 by Cavendish, who compared the gravitational attraction exerted upon a small pendulum by a large metal ball with that exerted by the Earth. This experiment allowed him to estimate that the density of the Earth was 5·45 gm/cc. Later, more accurate, experiments suggest that that figure is only fractionally low. This figure provides the basis for theories about the composition of the Earth.

It is easy enough to estimate the composition of the outermost layer of the Earth's crust simply by examining many different kinds of rock, estimating their relative common-ness, and averaging the result. Such an analysis tells us that 46·6 per cent of the crust is oxygen, 27·7 per cent is silicon, and an assortment of metals (aluminium, iron, calcium, sodium, potassium and magnesium in descending order of abundance) make up 24·3 per cent of the remainder. The rather insignificant remainder (about 1·5 per cent) includes hydrogen, nitrogen and phosphorus – all among the most important constituents of living matter – and carbon itself, the basic structural unit in the chemistry of life. Broadly speaking, therefore, the outermost layers of the object consist almost entirely of silica and silicates, with a few impurities.

In addition to this, we know from observation that aluminium silicates (e.g. granite) predominate in the continental masses, while magnesium silicates (e.g. basalt) are predominant in the deeper rocks exposed only beneath the oceans. The surface where life exists thus consists of granite rafts and salt-water puddles on a basalt base.

This analysis of the crust of the Earth gives an average density of 2·8 gm/cc – just over half the mean density of the Earth. This, in itself, does not argue that the core of the Earth is necessarily different in chemical composition from the outer rocks – the increased density might be the result of compression. However, though we cannot physically penetrate very deep into the Earth we do have ways of investigating its depths – primarily by studying the behaviour of waves conducted through it. These waves are manifested on the surface as earthquakes, and their progress through the Earth can be tracked by seismographs.

Just as sound changes with the density of its conductive medium (try listening to music with your head under water) so vibrations in the Earth change character as the density of rock increases, or when there is a discontinuity in the character of the rock. Study of seismic waves suggests that the density of the outer rocks does increase with depth down to 2,150 miles, where the density of the rocks is about 5·9 gm/cc. Below that there is a core where waves are not detectable. Some waves appear to pass across the core and reappear at the other side, but others are stopped dead and obliterated by it. This property is best explained by supposing that the core is liquid.

A little mathematical logic is necessary to complete the picture:

the mass of the outer layers is calculable and the mass of the core can be obtained by subtraction. Assuming that the compression of the core material follows a smoothly-increasing pattern then the outer density of the core is probably about 9·5 gm/cc and the density at the centre would be about 12 gm/cc. These figures conform with the suggestion that the core is liquid iron.

It seems unlikely that we will ever be able to verify by observation the theory that the Earth has a liquid iron core, but there is corroborative evidence of a kind. Lumps of rock frequently drift into the Earth's atmosphere from outer space and fall to Earth as meteorites. Analysis of those which actually reach the ground (most are so tiny that they burn up in the atmosphere thanks to frictional heat) shows that most are made either of iron (with a little nickel and cobalt) or of 'stone' (i.e. silicates *et al.*). Some contain both stony and ferrous material. The inference drawn from this is that if meteorites are made up of the same stuff that planets are made of, then the iron meteorites are 'core-substance', the stony ones 'mantle substance' and the occasional mixed ones are intermediate. It has been suggested that most meteorites come from the asteroid belt − an accumulation of debris orbiting the sun between Mars and Jupiter which may be the remains of a planet that disintegrated or the components of one that was never put together in the first place. The likelihood of this analogy's being sound depends very much on ·the competence of our assumption that planets are fairly similar − i.e. fairly Earthlike.

The Earth has three layers rather than two, although the discontinuity between the shallow crust and the mantle is less well defined seismically than that between mantle and core (it was discovered in 1909 by Mohorovicic and was named after him, although it is usually called the 'Moho' for short). There is at least a possibility of drilling through the crust into the mantle proper in order to confirm some of our suppositions because in some places (mostly at the bottom of the ocean) the crust is only three miles thick instead of the average thirty.

The point of all this is that the Earth is fairly complex as solid bodies go, and that what we know about its structure is to some extent tentative, partly based on faith in the uniformity of the planets. But the Earth is not just a solid body, and it is a grave mistake to think of it simply as a ball of rock. It has a very thin watery layer extending over most of its surface, which is chemically, if not structurally, complex − and whose presence at the solid/gas interface is vital to many of the things which go on at that interface. The gaseous component of the Earth-object is also much more complex, structurally if not chemically, than was generally realized until the present century.

Air, like water, is fluid, and the chemical composition of the atmosphere is fairly similar at top and bottom. In the atmosphere, unlike the solid part of the Earth, layers are demarcated principally by pressure and temperature.

It was at the turn of the century that high-altitude balloons first revealed that at a height of about six miles the steady decline in the temperature of the air stopped. This discontinuity was named the tropopause, and separates the troposphere from the stratosphere. About three-quarters of the mass of the atmosphere is probably in the troposphere. The stratosphere extends up to sixteen miles or so, and contains all but 2 per cent of the rest.

Immediately above the stratosphere is the mesosphere, and here the temperature rises as one ascends, then begins to fall again. Above the mesosphere is the extremely rarefied thermosphere, which contains only a very tiny (0·001 per cent) part of the atmosphere. Even here, however, the air is dense enough to heat meteors to incandescence so that they show as 'shooting stars'. Beyond the thermosphere − which extends to more than one hundred miles above the surface − is the exosphere, which is virtually a vacuum (and which gradually fades away to become the 'hard' vacuum of outer space) but which still contains enough matter for things to happen there: the auroral displays, for instance, take place in the exosphere.

These various layers of the atmosphere are primarily interesting − and structurally important − because of the effects which they have on the sun's radiation as it impinges on Earth. In the thermosphere there are layers of ions, stripped of their electrons by energetic photons and particles, which are more or less permanent, and which reflect radio-waves. The best known is the Heaviside layer, which is about seventy miles high. These ion layers may be disturbed by unusual solar activity.

In the mesosphere ultra-violet solar radiation is absorbed as 'fuel' for a reversible reaction converting oxygen molecules into ozone molecules, and this filtration of solar UV is most important in defining which component of the sun's radiation actually reaches the surface.

Lower down, at the top of the troposphere, there are two perpetual air-currents − the jet streams − moving east to west at several hundred miles per hour, which are important in determining conditions on the surface in terms of temperature and water-distribution (climate and weather).

Most of these effects of the atmospheric structure are vital only with respect to the organisation of the thin surface layer where life exists, but we should realise that the atmosphere is really a very complex physical structure, and no less a part of the Earth because it is not solid. We should, perhaps, also take note of the extent to which the nature of life at the surface – and perhaps the fact that it exists at all – depends on the properties of the atmospheric structure.

Having described briefly *what* the Earth is, we must now go on to ask *where* it is. It is, of course, the third planet of nine in a solar 'family'. But what kind of system is the solar system? Popular analogies attempt to communicate its scale and general aspect by comparing the sun to a beach ball and the planets to ball-bearings and grains of sand at various distances from it. I think, however, that it is misleading to represent the system as a collection of miscellaneous objects distributed in empty space, for the system as a whole has more integrity than that. It is also much more complex. It is true interplanetary space has so little matter in it that it is hardly there at all, but as we have seen with respect to the atmosphere, there are degrees of nothingness, and important events can still happen even in near-vacuum.

When we consider that interplanetary space is a harder vacuum than we can obtain in our laboratories at sea level, we are tempted to conceive of the Earth as a self-contained entity, separate from all else and quite independent except for the influx of solar radiation (as standardised, in our eyes, as the output from an electric fire or a light bulb). We tend, therefore, to underestimate the extent to which the Earth is influenced – and vulnerable to influence – by external forces. It is only recently, for instance, that we have come to realise how the weather on Earth may be affected by the weather in space.

The weather in space is, of course, produced by the sun, which is not just a gigantic light-bulb-cum-electric-fire in the sky, but an object just as complex as the Earth.

The first evidence which astronomers were able to use in order to draw conclusions about the nature of the sun was the behaviour of sunspots. The so-called 'sunspot-cycle' discovered in mid-nineteenth century by Schwabe (who was actually looking for something else entirely – a planet between Mercury and the sun) has long been one of the curiosities of science. The interval between peaks of sunspot activity is about eleven years.

The rotation of the sun was first discovered through sunspot observations, as was the fact that the sun is not solid, which was deduced in 1863 by Carrington, who observed that sunspots in different latitudes rotated with different periods. It was also Carrington who observed the first-recorded solar flare (and he might have gone on to further triumphs had he not elected to give up science in favour of managing the family brewery).

The apparent surface of the sun is, in fact, a structural layer called the photosphere, which produces the sun's visible light. Films of the photosphere reveal it to be a place of constant violent change, with 'granulations' in constant chaotic flux. Sunspots appear first as embryonic 'arch filaments' and grow rapidly to colossal size. The photosphere is a relatively shallow layer, about 200 miles deep, and the temperature at its surface is about 6,000 degrees Absolute.

The region outside the photosphere, which extends for several thousand miles, is the chromosphere, and beyond this is the corona which thins out gradually. A stream of particles – mostly electrons and protons – constantly expands out of the corona, and constitutes the 'solar wind'. These particles may be deflected by the Earth's magnetic field, and fall into tight spirals extending from pole to pole, known as the Van Allen belts. Where these belts dip deepest into the mesosphere of the Earth (at the poles) they cause the aurorae.

The corona of the sun consists of extremely hot (up to 2,000,000° Absolute) ionised gas. A gas in this state is called a plasma, and its behaviour is quite distinctive – for this reason plasma is often called the 'fourth state of matter' akin to but distinct from solid, liquid and gas. The temperature of the solar wind declines as it progresses into space, but in the neighbourhood of the Earth's orbit it is still about 100,000°. Although the density of the particles is only about 5–10 per cc this 'space weather' is by no means ineffectual.

The cycle of solar activity affects the solar wind, and is also correlated with magnetic storms. The statistical correlation was first observed soon after Schwabe's findings, when theory was quite inadequate to provide any kind of explanation – and ever since then there have been attempts to link all kinds of natural and pseudo-historical Earthly cycles to sunspots.

More recently, the observation that sunspots are not evenly distributed over the sun's surface and usually 'wander' longitudinally led to the idea that they may be influenced by planetary tides. The tidal effect of the Earth on the sun is very tiny (1/7000 of the force of the moon's tidal drag upon the Earth) and the tides caused by the other planets are of the same order of magnitude or less, but some degree of statistical correlation has been found between planetary conjunctions and solar activity. As events on Earth may thus be affected by conjunctions of the planets (via their affect on the sun and the solar wind) there has been some move to claim a 'scientific basis' for astrology. Had astrologers in antiquity built up their methodology by painstaking record-keeping and careful statistical sifting of data, there would be some substance to this claim, but as astrology was built on intuitive inspiration and mysticism it is really quite irrelevant. (But let us not underestimate the usefulness of intuitive inspiration, in science or in mysticism. It has a better record than statistics when it comes to finding out the truth.)

Astrological or not, the interaction between the planets and the sun is real. Being neither solid nor liquid – but very massive – the sun may be rather more amenable to tidal influence than the Earth's surface, and the relative smallness of the tidal force may not mean that it is altogether insignificant.

The Earth is not by any means the independent, self-contained entity that nineteenth-century scientists used to think. It is much more complex, and we can begin to see now just *how* complex is the balance of forces which determine conditions in the thin layer where we live. It is, in fact, a very special place indeed – though not in the same way that Aristotle and Aquinas once supposed.

The recent rediscovery of the uniqueness of the Earth has forced us to rethink many attractive ideas concerning the nature of the solar system of which we are a part, and of the universe as a whole. The strangeness of the universe – the essential alienness of the other worlds it contains – is only now becoming clear to us.

The closest and most easily observable object in the heavens is the moon. The idea that it was another world – an object akin to the Earth – first became popular knowledge in the seventeenth century, although even then it was a matter of some dispute. The status of the moon as a world was one of the two principal imaginative linch-pins of the new astronomy (the other being the fact that the Earth went round the sun and not vice versa) and both astronomers and philosophers of the new science made particular attempts to popularise the notion. John Wilkins, one of the leading English scholars of the age, wrote a treatise called *The Discovery of a New World in the Moon* earnestly arguing the point; and Johannes Kepler, imperial mathematician to the Holy Roman Empire – the man who first described the planetary orbits – made an even bolder attempt at popularisation when he turned his *Lunar Astronomy* into a piece of imaginative fiction, the *Somnium*, which gives an account of the observations made by a man carried to the moon in a dream. The passion to persuade led both these men into new areas of speculation: Wilkins was the first man to suggest seriously that men might one day travel to the moon (and thus prove all his assertions), while Kepler took it upon himself to design a fauna appropriate to conditions upon the moon as he envisaged them.

The accounts of the moon as a world given by Wilkins and Kepler are inordinately perspicacious – both the *Discovery* and the *Somnium* are excellent examples of the seventeenth-century scientific imagination. But Wilkins and Kepler – and most of their contemporaries – were wrong in several important respects. All the mistakes they made derived from the one basic assumption that the moon had more in common with the Earth than proved to be the case. They both credited the moon with an atmosphere and with surface water. They pictured the moon, in fact, by imagining a smaller Earth with a longer day/night cycle.

Time, however, has shown that the moon is not at all the same kind of object as the Earth. It has hardly any gaseous component at all – certainly not an atmospheric structure as complex as the Earth's. Its mean density is about $3 \cdot 3$ gm/cc – about the same as that of the Earth's crust plus mantle – and thus presumably has a rather different recipe even so far as its solid component is concerned (moon rock brought back by the Apollo astronauts has proved to be rock, not very different from the rock which comprises the Earth's outermost solid layer, but that is only scratching the surface). Spacecraft making low passes over the lunar surface have detected the presence of concentrations of denser material below, but fairly close to the lunar surface. These have been dubbed *mascons*, and may be composed of matter more like that which is supposed to comprise the Earth's core – i.e. iron – but which may have been introduced from without by meteor-strike. The craters on the moon are, of course, testimony to the vulnerability of the moon to meteor impact – without an insulating atmosphere there is nothing to protect the moon from meteors, and there is no erosion to smooth away the scars once they appear.

The moon is not completely dead internally – there are occasional moonquakes and some volcanic activity, but these are not nearly so frequent or so impressive as the shocks suffered by the turbulent Earth. Just about the worst quakes the moon has suffered lately happened when the lunar ascent stages of the Apollo landing-craft were sent crashing back to the surface.

The moon appears to be little more than an inert lump of stone, but it is important as an agent of Earthly change – and here again we have an instance of the extent to which conditions on Earth depend on its integrity with the solar system as a whole. Tides affect the Earth's solid surface as well as the oceans, and the pull of the moon is gradually slowing down the Earth's rotation upon its axis. The day is presently gaining about one second per fifty thousand years. The day in Devonian times, 350,000,000 years ago, was only about twenty-two hours long, and there were four hundred days in a year. (Fossil corals from the Devonian age have daily growth rings as well as annual ones, and correspond well with the figure calculated theoretically.) In about 3,000,000 years our descendants (if any) will be able to dispense with leap years. The moon, of course, has been so affected by the Earth's tidal drag that its day is now as long as its 'year', and it keeps the same face perpetually towards us. It is also very gradually spiralling away from us, so that the lunar month is also increasing, though not as fast as the Earthly day: it will take about 10,000,000,000 years before the two are equal and the Earth and moon will present the same face perpetually to one another. Even this will not mark the end of the delicate relationship – solar tides will continue to decrease the length of the Earthly day, and the lunar tide will then act in opposition to it. This will cause the moon to begin spiralling inward again. Mathematical models of the situation suggest that there will be no ultimate equilibrium and that ultimately the moon will come so close that it breaks up under gravitational stress – the Earth may then become a ringed planet, like Saturn.

Speculation as to how the relationship between Earth and moon first began encounters logical difficulties in every direction, and provides something of a mystery. The simplest suggestion is that the Earth and the moon formed together during the origin of the solar system, but it is very difficult to imagine how one loose aggregation of mass could coalesce into two large spherical bodies. On the other hand, the moon is rather a large object to have been captured. Those whose imagination favours spectacular explanations have suggested that the moon may have been ripped out of the Earth's corpus by cosmic disaster of one kind or another, but this is one idea which does not seem competent in the light of comparison between moon rock and the rocks of the Pacific Ocean, from whence it is supposed to have come.

It is interesting to note here for the first time a polarisation of imaginative explanation which we shall encounter again: the choice between catastrophist explanations and uniformitarian ones. When constructing imaginative models to explain how a state of affairs came into being, people seem to have an intellectual affinity with either one kind of thinking or the other. Catastrophists believe in bold explanations which throw all enigmas into an ideative cooking-pot and concoct a single sequence of violent events. Uniformitarians believe that all change is very slow, and comes about through the elaborate and delicate collaboration of many unrelated events. The arguments which arise whenever the two schools of thought compete for the privilege of explaining an enigma is invariably fierce, and suggests strongly that the issue goes much deeper than intellectual expertise, involving emotional reactions and committed beliefs. Both sides have had their victories – the uniformitarians were triumphant in the argument about the age of the Earth and the process of evolution which raged throughout the nineteenth century, but the catastrophists won out convincingly in the twentieth-century debate between the big bang theory of cosmogenesis and the steady-state model of the universe. In the case of the Earth–moon relationship the evidence of the fossil corals comes down strongly in favour of the uniformitarian side, but the fact that there is a sharp break in the fossil record at the beginning of the Cambrian period, when the Earth was apparently scrubbed clean by some momentous process, leaves room for some lines of catastrophist thinking. Incidentally, there is no one more willing to state his opinions as 'scientific facts' than a committed catastrophist or uniformitarian, and this applies to *both* orthodox scientists and 'scientific heretics'.

The mistake which Wilkins and Kepler made concerning the moon was carried into the twentieth century by scientists attempting to guess what conditions on other planets might be like. This is how the clichéd images of Venus and Mars, which were so completely destroyed by the space-probes of the 1960s, first arose.

Optical astronomers knew just three things about Venus – its size, its mass (calculated from its gravitational effect) and the fact that it had a high albedo (that is to say, it reflected a high proportion of the sun's light). Its mass and size were not so different from the Earth's, its diameter being a couple of hundred miles less and its mean density $4·9$ gm/cc, as opposed to $5·5$ for the Earth. It was assumed that it reflected a lot of light because its atmosphere was cloudy – no surface features could be distinguished.

Scientists tried to imagine what Venus must be like by imagining what conditions on Earth would be like if it were as close to the sun as Venus, and had a lot more clouds. They came up with a warm (about $65°$ C) planet with lots of water. Science fiction writers habitually translated this into a planet of steaming jungles or a world completely covered by an ocean.

Despite the similarity of size and mass between Venus and the Earth, however, it turned out that Venus was not at all the same kind of object, and that in fact conditions were very different from what conditions on Earth might be if the Earth were in Venus's orbit. The surface temperature is, in fact, nearer six hundred degrees than sixty, and the atmosphere is structurally very dissimilar to the Earth's. It seems, in fact, to contain several hundred times as much mass as the Earth's. It is mostly carbon dioxide, and the extreme surface temperature may be caused by the 'greenhouse effect' characteristically associated with that gas – high-energy photons from the sun go through the atmosphere and heat up the surface, but the energy which would be re-radiated as infra-red light is absorbed by the atmosphere and does not escape. There is no nitrogen in Venus's atmosphere and no oxygen. There is no spectroscopic evidence of water vapour, but the Soviet Venera probes which have descended into the atmosphere found large quantities of it in the lower regions. The atmospheric structure may therefore be complex. The clouds may be sulphuric acid, and there are almost certainly other corrosive acids present in the upper regions. So far as we can ascertain, Venus is very different indeed from the world our Earth-centred assumptions led us to believe in.

By the same process of analogy, Mars became, in the

nineteenth-century imagination, the planet that the Earth would be if it were at the appropriate distance from the sun (although physically, Mars is more akin to the moon than the Earth, having a diameter about half the Earth's and a mean density of 3·95 gm/cc). It was credited with a rarefied nitrogen/oxygen atmosphere and a surface temperature near to the freezing-point of water. Unlike Venus, however, Mars actually had features visible to optical astronomers. The optical astronomers thus had an additional advantage in trying to deduce what conditions there were like. Or did they?

In actual fact, the answer appears to have been 'No'. Far from the sight of the surface being an advantage to observers, it was a seductive trap, for it made them prey to all the illusions which sight permits. (And it should be remembered that even the best telescopes give an image of Mars much less clear than the face of the moon as seen with the naked eye.)

It was in 1877 that the popular image of Mars as an Earthlike world mostly made up of red desert gained tremendous imaginative support when Schiaparelli saw lines crisscrossing the face of the planet and named them *cañali* (translated as canals, though more accurately channels). Photographic techniques in Schiaparelli's day were not well-developed, and all the accounts of the canals of Mars were produced by astronomers who drew them. What they drew, of course, depended upon what they saw, and what they saw depended very much on what they believed. Men with an intellectual affinity for the idea of Martians could practically see the barges and the docks. (Seeing is, to a very large extent, an interpretative process. The information which we actually *use* in making a mental image out of an image on the retina is very largely supplied by the brain rather than the eye, and we actually do see very much what we *expect* to see. Kittens reared in an environment where everything is vertically striped cannot, without re-training, distinguish between horizontal stripes and series of dots. The men who saw the Martian canals were *not* incompetent observers, merely imaginative ones ... and if all scientists rigorously resisted seeing anything new in what they observed we would have had very little progress in science at all.)

As time went by, the old red-desert Mars, pictured so eloquently by the American astronomer Percival Lowell and romanticised by Edgar Rice Burroughs, went out of fashion — although it is still remembered with considerable imaginative nostalgia. The Mariner spacecraft which have been visiting Mars for some time now have revealed a surface rather like that of the moon – pock-marked with craters. The atmosphere is extremely tenuous, containing only slight traces of oxygen and virtually no water. The polar ice-caps are extremely cold ice. The hope that Mars might harbour life of a sort has been abandoned save where it survives as a relic of a hope once so much fonder.

And so the catalogue continues. The more we come to know about the planets the more we realise how different from our own they must be. Jupiter and Saturn, like Venus, have opaque atmospheres of unknown depth, and what lies beneath them remains very much an open question. Such information as we have, brought back by spacecraft which have recorded radio signals from within the Jovian atmosphere, is vague and uncertain. Speculation about the kinds of objects which the outer planets may be is necessarily very tentative. Several present fascinating enigmas – the Great Red Spot of Jupiter, the extraordinary apparent density of Pluto, the fact that Uranus is tilted so far over on its axis that it is rolling around in orbit on its side – to which we have as yet only speculative answers. The outer planets are also well-supplied with satellites (the inner ones are not – the Earth–moon system is really a 'binary planet' and Mars's two satellites are extremely small). Pioneer 10, which by-passed Jupiter in 1973, confirmed that at least one of Jupiter's satellites (Io) has a tenuous atmosphere, and it seems probable that some of Saturn's moons are also objects as structurally complex as the planets themselves. Titan, for instance, may have a moderately dense atmosphere.

There is no room here for an elaborate catalogue of the planets and their attendant satellites, and such a list would not by any means complete a full account of the population of the solar system. There still remain countless small planetoids, most of which orbit in the 'asteroid belt', and a large population of comets, most of which appear only briefly and unostentatiously within the reach of our telescopes.

For many centuries astronomers could not explain either comets or meteors – some exceptionally hard-headed astronomers of the nineteenth century were still denying the very existence of meteors, believing them to be hallucinatory, and it was certainly not generally accepted that they were of extra-terrestrial origin. Comets, apparently random in their heavenly visitations, were, like many other unpredictable aspects of nature, correlated with events which upset the order of human life. When William Miller predicted in 1843 that the world would end on 21 March a new comet appearing in the sky shortly before that date lent such power to his prophecy that the expiration of the ultimatum could not shake the faith of the committed. The Church which Miller founded split into two and both halves are still flourishing today – and still expecting the imminent end of the world.

The posthumous confirmation of Halley's prediction concerning the return of the comet which now bears his name (in 1758) first allowed mathematical reasoning to begin the work of rationalising comets, but it was not until much later that they were explained as agglomerations of 'ices' (water, methane, carbon dioxide etc.) which vaporise as the comet sails into the solar wind and stream out behind in the well-known 'tail'. (In 1910 the Earth passed through the tail of Halley's comet without any noticeable effect.) Comets appear to have a limited lifetime, becoming successively less bright with each reappearance – and this implies that there must be a constant supply of new ones, suggesting that there may be a good deal of debris beyond Pluto. There is more to the solar system than we can see, and much more than we are ready to assume on the basis of what we *do* see.

There is a certain irony in the fact that twentieth-century science has given us elaborate and apparently competent models of the atom and the universe, but has not been able to present us with a good account of the nature, organisation and origin of the solar system. Perhaps this is testimony to the fact that things are generally more complicated than we imagine; perhaps we only understand the atom and the universe so well because we have so little direct knowledge of them – the snags and the flaws do not show up because we do not see them. The solar system, being neither too small nor too vast to perceive directly, is a more awkward thing to deal with. If the solar system, like the atom and the universe, were an imaginative construction, designed according to the specifications of mathematical aesthetics, who would have cluttered it up with so much debris? Who would have made the planets such different sizes and used such different recipes? Who would have put a red spot on Jupiter and let Uranus fall over in its orbit? Who would have allowed the sun to grow blemishes continually?

The difficulties which we have in constructing a theory to account for the solar system are difficulties caused by a superabundance of information. This is why it is in this realm, particularly, that new discoveries tend to reveal such embarrassing self-deceptions entertained in the past. In atomic theory progress tends to consist of old misunderstandings being cleared up by brilliant new insights. In astronomy old understandings are smashed up by the revelation of prior oversights. No brilliant new synthesis is emerging to account for it all: merely an ever-extending catalogue of exceptions and anomalies.

Kepler, who was the first man to show that the planetary orbits are elliptical and that the square of the time a planet takes to complete an orbit is proportional to the cube of its distance from the sun, also tried to find a mathematical key to the pattern of the orbits. What he was looking for was a kind of astronomical equivalent of Balmer's Ladder: a series which would reveal the 'divine plan' to which the solar system had been constructed. He failed. In 1772, however, Titius found a mathematical series which seemed to bear some relationship to the orbits of such planets as were known at that time. This series was popularised by Bode and became known as Bode's Law.

The basic series is 3,6,12,24 .... Prefacing the series with a nought and then adding 4 to each element gives the sequence 4,7,10,16,28 etc. The mean distances of the various planets could be fitted to this sequence if the Earth's mean distance was standardized as 10. The calculated value for Mercury was 3·9, for Venus 7, for Mars 15, for Jupiter 52 (exactly right) and for Saturn 95 (as opposed to 100). The errors were not large, but one figure was missing from the planetary sequence: 28. Bode's supporters began looking for a new planet to fit the figure, and in 1801 they found one – Ceres. Within the next few years several more – all, like Ceres, very small – were located, with approximately the same mean distance from the sun.

In the meantime, Herschel had discovered Uranus in 1781 (at first mistaking it for a comet) and it, too, had turned out to fit the Bode sequence, with a mean orbital distance of 192 on the Earth–10 scale, compared to a Bode sequence number of 196.

It seemed, therefore, that the logical plan of the solar system had been discovered. The problem was to do for Bode's Law what Bohr had done for Balmer's Ladder and translate the mathematical idea into theory by discovering its physical cause. But the situation was yet to become more complicated. According to Bode's Law an eighth planet (if there was one) ought to be at

388 on the scale. Measurements of the orbital perturbation of Uranus confirmed that there had to be such an eighth planet to exert the gravitational attraction which could not be accounted for with reference to the known planets. But when the eighth planet did turn up it was not in the right place − nor was it massive enough to account for Uranus's wandering. It was, in fact, at 301 on the Earth−10 scale. Then came Pluto, which at 394 on the Earth−10 mean distance scale, was more or less where Neptune 'should' have been. This far out, however, the orbits are much more eccentric ellipses, and Pluto's orbit actually crosses Neptune's, so that some of the time Pluto is the eighth planet and Neptune the ninth.

Can Neptune and Pluto be lumped together (like the asteroids) for the convenience of Bode's Law? Alternatively, can one of them be condemned as an 'invader' which has upset the natural beauty of the Bode sequence in being captured by the sun? Well, perhaps. Opponents of Bode's Law claim that fitting a numerical sequence to eight different distances is not such a difficult task, and that whatever the distances were *some* series would provide a fairly close fit. Those who favour the pattern would counter this with the argument that a sequence derived to fit five planets which actually turns out to fit seven plus an asteroid belt must have more to it than coincidence − it must be a *rule* rather than a *description*. But at the moment, Bode's Law *is* just a description, and a slightly inaccurate one at that, because we have not yet managed to explain it in terms of how the solar system came to be as it is in the first place.

As in the case of the Earth−moon relationship, every speculative account of the origin of the solar system can be met with logical objections.

The fact that the planets are distributed in a single plane is highly suggestive. It is the plane of the sun's own rotation, and suggests that while the sun was rather more loosely aggregated than it is now the inertia of its rotation may have caused smaller lumps of matter to fly off and distribute themselves as the solar family. A variant of this idea is that as the sun was condensing from a cloud of gas, various 'currents' or 'vortices' were set up in the cloud and resulted in the separate condensation of matter into several planetary bodies, whose distribution is thus governed by the initial vortex pattern.

The principal alternatives to this kind of 'centrifugal' theory are provided by various catastrophist models in which the sun, while condensing or afterwards, passed close to or collided with some other body, and the resultant forces spewed matter out all over the place. The planets may then be bits of the sun or bits of the other body or bits of both. With the cataclysmic chaos of a cosmic disaster it is easier to explain the fact that the asteroids either never became a planet or are the remains of a shattered one, and the general untidiness of the whole solar system begins to seem less problematic. Actually writing the script for such a catastrophe is, however, very hard to do, because when each anomaly has to be explained separately in relation to the same cause the credibility of the event is gradually stretched to breaking-point. Uniformitarians, taking advantage of the great age of the solar system, have ample opportunity to introduce the anomalies one by one as a result of slow processes and singular events.

The chief objection to the centrifugal hypothesis is provided by the very different chemical composition of the planets: if they were all formed at much the same time by much the same process, why does the Earth seem to be mostly oxygen and iron while Jupiter is almost all hydrogen? This objection is partly overcome by suggesting that the planets all started out pretty much alike, but have since changed because of the solar wind, which carries lots of stray hydrogen and helium out of the sun, and distributes most of it in the regions where the gas giants gradually accumulate it by gravitational attraction. But there is also the question of Uranus, which rotates on its own axis at right angles to the ecliptic – how can this be if the initial impetus given to the planets was more or less the same? Perhaps we must introduce just a *little* catastrophe to knock Uranus about a bit, but once we have catastrophe, why not extend it to the whole solar system ...?

At present, it seems that the 'vortex' theory might offer the most hope of finding a model to explain the basic consistency within which the troublesome anomalies occur. It may also be that an explanation of Bode's Law can be found by considering the effects which the planets have on each other rather than simply their relationship with the sun. In any case, any understanding to which we might lay claim is necessarily very tentative. The fact is that we simply do not know.

The point that I have tried to make in this chapter is this: that discovery and self-deception go hand in hand. The keys to understanding that we have tried to use in unlocking the secrets of the solar system have proved inadequate because of our readiness to assume that what holds for the Earth will hold elsewhere. The twentieth century has given us new tools of observation (radio-astronomy and space-probes) with which we have sought to confirm the evidence of our eyes and instead make a mockery of it. We know better now – but how much better? The alternative routes to understanding – the rationalisation of mathematical distributions – which have served us so well in other areas of science have not served us nearly so well here, perhaps because we have so much more than the numbers to confuse us.

New discoveries reveal old self-deceptions, but *they do not remove the source of such self-deception*: the egocentric illusion that the lessons of our experience (whether it be sensory experience or mathematical experience) can be extended to apply to the universe as a whole. Our assumptions betray us, again and again. There is at least a possibility that the knowledge which we have of the atomic microcosm is so competent and so assured simply because we are not in a position to detect the exceptions and the anomalies which upset its logic: they are lost within the pattern which dominates and which – by statistical aggregation – confers its properties on the observable world. While we cannot explain what we *can* see, and are so often deluded as to what we *do* see, how can we be so sure of our triumphs in understanding that which we cannot?

4

The Expanding Universe

The stars, like the planets, belong to the visible universe, and the first steps in understanding their nature − and the nature of the macrocosm whose 'atoms' they are − were taken by optical astronomers. With respect to the stars, however, the information obtainable *via* optical astronomy is very restricted. Several of the planets presented visible faces − measurable discs − to simple telescopes, but the stars are only points of light; when Galileo first made observations with his telescope the only thing he learned about the stars was that there are far more of them than are accessible to the naked eye. In addition, the stars appear, at first, to manifest no behaviour − they were known even during the Renaissance as the 'fixed stars', because they changed neither their position nor their aspect. Not until 1572, when a 'new star' appeared in the sky and was studied over a period of time by Tycho Brahe, was faith in the Aristotelian notion of the stars as a kind of 'cosmic wallpaper' occupying the outermost sphere of the universe seriously challenged (despite the fact that there had been a much brighter 'new star' − like Tycho's star, a supernova − in 1054, observed and recorded by the Chinese but ignored in the Christian world).

The idea that our sun is, in fact, a star, and that there might be an infinite universe filled with such stars, was hinted at by Nicholas of Cusa in the fifteenth century and taken up by the ill-fated Giordano Bruno at the end of the sixteenth, but it really remained a matter of pure speculation for at least a century after that. Not until 1718 did Halley offer convincing evidence of the fact that the stars *did* move relative to one another when he compared his measurement of their various positions with those of Hipparchus (in 134 BC). This was the first important indication that the stars conducted their business on a time-scale very different from our own − the first indication, in fact, that there *were* different time-scales appropriate to different classes of entities.

Attempts to discover how far away the stars were from our vantage point in space met with little initial success. Huygens, in the late seventeenth century, had tried to estimate their distance by assuming that they were suns like our sun, of comparable luminosity, and though this very rough estimate actually gave something like the right distance-scale, the assumptions were too shaky to give the answer much conviction. The distances of the planets had been calculated by parallax (that is to say, measuring their position relative to the stellar background from two different places, thus achieving 'binocular vision' – the further away the planet was, the less it seemed to 'move' relative to the background). Measuring the parallax of a star was, however, a rather different matter – no movement relative to a background could be detected for any star during the seventeenth and eighteenth centuries. Not until the last years of the 1840s, when Bessel, Henderson and Struve independently measured the parallaxes of 61 Cygni, Alpha Centauri and Vega respectively, was the distance-scale of the stellar universe first revealed. Alpha Centauri, the nearest of the three stars whose distances were calculated, proved to be 25,000,000,000,000 miles away – so far that a new unit of measurement was needed to make the figures conveniently modest. The first such unit in common use was the parsec – the distance at which the radius of the Earth's orbit would correspond to one second of parallax – but it has since been replaced by the light-year, the distance travelled by light in a year.

Once this vital initial measurement was made, the whole situation became much clearer. It was known that the stars were distributed in a vast volume of space, and that their luminosity was not related to distance alone. At this time the notion of an infinite universe filled with stars did not seem logically competent because of Olbers' Paradox which pointed out that if there were an infinite number of stars evenly distributed in space, the Earth ought to be receiving an infinite amount of starlight. It was thought that the lens-shaped 'sidereal system' supposed by Herschel and Kapteyn (i.e. the galaxy) comprised the whole of a finite universe.

This, essentially, remained the situation throughout the rest of the nineteenth century. New observations accumulated, but nothing happened to change our basic concept of what the macrocosm was like. Curiously enough, the imaginative instrument which was, during the twentieth century, to allow astronomers to discover a whole new macrocosmic concept, was discovered in 1842, but the problems to which its application was to prove crucial did not crop up until 1912.

The imaginative instrument so vital to twentieth-century measurements of the universe is the Doppler effect. This was first proposed by Doppler to explain colour differences between components of binary stars (differences which later turned out to have nothing to do with the Doppler effect). Doppler pointed out that the wavelength of light emitted by a moving source may be affected by the movement − a source moving towards the observer would emit slightly compressed light-waves, whose wavelength would be decreased, while a source moving away would emit light-waves which were slightly stretched, and whose wavelength would thus be increased. Because stars emit light over a wide range of the electromagnetic spectrum the colour of the light emitted by a moving star (despite what Doppler thought) would not normally be affected. What would be affected, however, is the hydrogen signature of the spectrum: the specific lines characteristic of incandescent hydrogen would be shifted slightly because of the compression or stretching of the emitted light. If the star were moving away, the lines would be shifted towards the low-frequency end of the spectrum − that is to say, 'red-ward'; whereas if it were moving towards the observer it would be shifted 'violet-ward', towards the high-frequency end of the spectrum. It is through these shifts (particularly the red-shift) that the Doppler effect has become so important in telling us what is happening in the macrocosm.

The information provided by measurement of red-shifts is, in itself, very slight − it tells us whether the distance between ourselves and another star is increasing or decreasing. But we can also detect, by optical observation, the motion of a star *across* our field of vision − at right-angles to the direction revealed by the Doppler shift. Now, any particular star might be travelling directly away from or towards us or, on the other hand, it may be moving 'sideways' relative to us, so that it never gets any nearer or any further away. If, however, we take the average of a number of stars, we may expect these relative motions to average out so that velocities towards/away from us are roughly equal to velocities sideways-on to us. In this way, the averaging of the Doppler shifts of a large number of stars which are associated with one another in a cluster of some kind will give us a reliable estimate of the sort of velocities they are liable to possess across our field of vision. Given the velocity and the angular displacement observed in optical telescopes, we can then calculate how far away they are.

This was first done in 1913, when it became necessary to 'calibrate' a new yardstick for measuring distance in the macrocosm – the yardstick provided by the 'Cepheid variables'. The Cepheid variables are a class of stars which vary in luminosity in a very regular manner. In 1912, Henrietta Leavitt had located and intensively studied hundreds of such variables in a star cluster called the Lesser Magellanic Cloud. She had discovered that the period of their luminosity-cycle (i.e. the time interval between peaks) was related to their magnitude (i.e. to their brightness at peak). It could be reasonably assumed that all the stars in the cluster were at much the same distance relative to the Earth, but there was no way of knowing just how far that was. The period/luminosity relationship of the Cepheids offered a method of finding out – if only we knew the distance of a single Cepheid anywhere in the known universe. Unfortunately, we did not – and had it not been for the method of averaging Doppler shifts we might not have been able to find the one vital reference point which gave optical astronomers the key to the Cepheid yardstick – and thus to the measurement of distances in the macrocosm.

The breakthrough was decisive in proving that the Herschel–Kapteyn sidereal system was not all there was to the universe. A whole host of globular clusters of stars were found to be outside the galactic lens. The Lesser Magellanic Cloud, its greater counterpart, and a number of other nebular clusters proved to be 'satellite galaxies' related to our galactic lens but not inside it. And beyond this handful of galactic neighbours, there proved to be a vast number of other galaxies and galactic clusters, extending, so it seemed, *ad infinitum*.

The conceptual size of man's universe thus took another great leap. Just as the enclosed Earth had given way to the enclosed solar system, and then to the sidereal system, so the sidereal system had to give way to a new and greater system whose units were not stars, but galactic organisations of stars by the billion.

The discovery of a new and vaster macrocosm was not the only discovery which Doppler shifts helped astronomers to make in 1912 and the years which followed, but I want to remain, for a while, with the stars rather than the galaxies, and consider the time-scale of the macrocosm with reference to the events which happen in the lifetime of a star.

Herschel once compared the task of an astronomer in attempting to discover the secrets of the lives of the stars to the task of a city-dweller allowed to spend a few minutes in a forest and then required to give an account of the life-history of a tree. In looking at the sky we cannot see stars 'being born' or 'growing old' because these things happen over a period of time within which our years and centuries of observation are negligible. What we see is essentially a 'still' picture which may contain stars at every stage in their evolution but which offers very little direct evidence of the processes by which they pass from one stage to another. Very occasionally, cataclysmic events may reveal themselves to us – although only the most cataclysmic of all will actually attract our attention and make themselves obvious. We have records of only three supernovae within the last thousand years, and the last was observed in 1604.

The rarity (on our time-scale) of observable *events* is only part of the problem. There is also the problem of unobservable *objects*. Objects within the solar system may be revealed to us by the reflected light of the sun or by the gravitational effects which they have on observable objects. Beyond the solar system, however, objects can only be revealed to us if they actually emit radiation, or if they are close enough to an easily-observed object to disturb its behaviour measurably. Optical astronomers have detected a host of 'dark stars' which are invisible companions of bright stars thanks to eclipses of the bright stars by the dark; but of dark stars alone in space we can know nothing. In the immediate neighbourhood of the sun – within fifteen light-years or so – there *appear* to be thirty to forty stars, but we have no way of knowing whether there might actually be twice that number or ten times as many. We cannot count the stars we cannot see.

The dimmest star on record is Proxima Centauri – dark companion to the visible binary Alpha Centauri. It is, not unnaturally, also the closest star known.

The luminosity of a star is related to its mass, but not linearly. A star ten times as massive as the sun emits about a thousand times as much radiation, and a star thirty times as massive burns a hundred thousand times as brightly. This works both ways – 61 Cygni A, whose mass is only 30 per cent less than the sun's, is one tenth as bright. A star one tenth the mass of the sun would emit less than one thousandth of the sun's radiation. In addition, a star much more luminous than the sun might be expected to be rather less stable and considerably shorter-lived. This applies the other way round, too. So, for every great, bright star that decorates the heavens, there must be countless small, feeble ones beyond the power of our telescopes.

Thus, in trying to piece together the life-story of a star from the pictures of individuals revealed by our telescopes, we must remember that our record is incomplete. We may be in the same situation as an alien trying to understand the human life-cycle on the basis of photographs of the population of a nursery school or an old people's home.

The sheer variety of the stars in the sky was not appreciated until 1881, when Michelson (co-proponent of the notorious ether-proving/disproving experiment) invented the interferometer. This device made it possible to measure the diameter of a star (an optically impossible task) by measuring the interference between rays of light emanating from the two fringes of the sun. It was not actually used for this purpose until the twentieth century, after Hertzsprung had deduced that certain bright stars whose spectra suggested that they were very cool relative to the stellar average, had to be very large. In pursuit of this suggestion, it was found in 1920 that the star Betelgeuse had a diameter 350 times the diameter of the sun and was thus a 'giant' – specifically, by virtue of its spectral class, a *red giant*. Some giant stars akin to Betelgeuse are so cool that they radiate almost all their energy in the infra-red range, and are therefore invisible, though detectable by infra-red photography. These infra-red giants may have seven or eight times the diameter of Betelgeuse.

The attempt to find some sense in the distribution of stars among the 'spectral classes' led Hertzsprung (and, independently, another astronomer named Russell) to draw a graph plotting the absolute magnitude of stars (i.e. their intrinsic luminosity as opposed to their brightness as perceived from the Earth) against their spectral type in order of descending temperature. The kind of result they obtained is shown in Figure 3. Most of the stars fall into the long curve described as the 'main sequence'. The giants fall into a broad 'branch' above the main sequence, except for a few supergiants right at the top of the diagram. Some of the dwarf stars (the red dwarfs) fill the 'tail' of the main sequence, but others (the 'sub-dwarfs') form a rough line parallel to the mid-section of the main sequence. The few white dwarfs scattered at the bottom of the diagram are not representative of the true numbers of dwarf stars which exist at the lowest magnitudes because most stars at these absolute magnitudes are invisible from Earth.

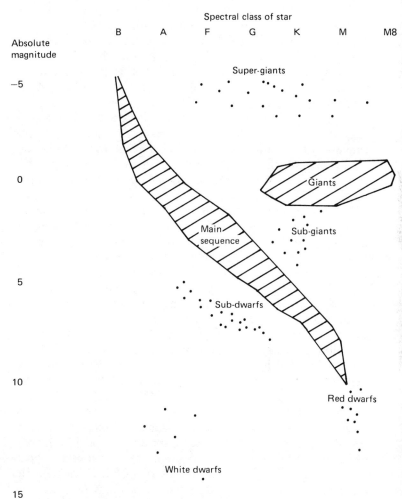

Figure 3   The Hertzsprung-Russell Diagram

The Hertzsprung-Russell diagram has become the frame of reference of the construction of theories of stellar evolution (much as the hydrogen signature became the pattern which had to be explained by theories of the atom). The various biographies which have been concocted are necessarily tentative, but the basic pattern of change is now reasonably well understood.

The density of stars in the universe is not very great, and there is a vast amount of empty space between them, but emptiness is only relative. The empty space between stars contains less than one millionth of the mass per unit volume that a man-made vacuum contains, and matter in the intergalactic vacuum is a thousand times less dense than that. The density of intergalactic matter is about one hydrogen atom per cubic centimetre. There is, however, an awful lot of intergalactic space, and calculations suggest that the intergalactic vacuum in the known universe contains several hundred times as much matter as all the galaxies put together. By the same token, the great bulk of a galaxy's mass is distributed in space – only a tiny fraction of it is formed into stars.

But this matter is not evenly distributed in space – it aggregates into clouds of various sizes and densities. The more it aggregates the more it tends to aggregate because of gravity (the galaxies are themselves clouds, which contain denser clouds, and so on). The more that a gas-cloud coalesces, the faster the atoms of hydrogen will fall towards its centre, and the more frequent will become the collisions between the atoms. As mass accumulates it will also begin to absorb more radiant energy, and the temperature of the cloud increases. This process continues for a few million years, and ultimately the temperature at the core becomes sufficient to permit fusion of hydrogen into helium – thus 'igniting' a star. Before this happens the cloud may have been able to shine feebly by emitting some energy from a large 'proto-star' surface.

Once the ignition of the star has occurred the initial process of change gives way to a kind of balance: further gravitational condensation is opposed by the expansion prompted by energy-production at the core. It is at this point that the star becomes part of the main sequence of the Hertzsprung-Russell diagram. Its actual station on the curve of the main sequence is determined by its mass – the greater the gravitational force within the star, the brighter it has to burn in order to stay in balance.

Main-sequence hydrogen-burning stars shine with a steady light over a long period of time – any fluctuation being compensated by feedback in the gravitation/energy balance. If the star's coalescence is represented as its 'childhood', and ignition as its 'adolescence', then its time on the main sequence is its 'working life'.

But not all stars actually reach the temperature necessary to ignite them − or so theory suggests. A proto-star with a mass less than 8 per cent of the sun's would probably never generate sufficient heat to begin fusion, and in this case gravitational collapse would render the gas in the cloud 'degenerate' (in 'degenerate matter' increasing compression and density no longer result in increasing temperature) and it would end up as a black dwarf composed of 'collapsed matter'.

As the main-sequence star advances in age towards senility, more and more of its hydrogen is transmuted by fusion into helium. Helium is heavier than hydrogen, and thus the helium tends to become concentrated in the core of the star. The older the star gets the larger the helium core becomes, and the more tightly it gets packed. The nuclear reaction balancing out the gravitational forces is peripheral to this core, and in order to maintain the gravitational balance of the whole star against the increasing concentration of mass at the core the star expands to become a 'sub-giant'. Stars of very large mass indeed may become 'super-giants' at this stage in their development.

But the situation is now rapidly becoming unbalanced. The internal temperature of the star continues to increase until it crosses a critical threshold and the balance between contraction and expansion can no longer be maintained. The core becomes hot enough for helium to begin fusing into elements of progressively greater atomic weight, and as each 'fusion-stage' is passed the energy-liberation of the star changes. Instead of being stable the energy/gravitation balance may become cyclic − and sometimes very regular, as in the case of the Cepheid variables.

When a star's light is least stable it tends to have reached the limits of its potential luminosity, and with the energy-producing potential of the core now declining, gravitation begins to get the upper hand in the long struggle. The stars may blow off mass in a series of 'core pops' − minor explosions which temporarily release enough energy to counter collapse − or even throw off the greater part of their mass in one vast explosion: a supernova. Either way, the star will ultimately work its way to the same final fate which befell such stars as never ignited in the first place: they will lose most of their mass and all of their energy-producing capacity and end up an ember: a dwarf star of degenerate matter. The degree of collapse, and the potential of the ember to radiate such energy as it has left, depend very much on the route which the star takes to this final resting place. Supernova relics may degenerate not only to 'collapsed matter' but to 'neutron star' status, when the electrons of the atoms are forced into the nuclei, and combine with the protons to make neutrons, thus turning the matter into pure 'neutronium'.

But this is not quite the end of the story, because the mass which is blown off from the aging stars is returned to interstellar space. This mass comprises not just 'primal' protons and electrons – the hydrogen constituents from which the star first condensed – but ions and atoms of elements of much higher atomic weight. Thus, to the clouds of interstellar gas will be added clouds of interstellar *dust*. Then, a 'second-generation' star, whose initial composition is partly made up of material from an already-dead star, may form. A second-generation star will move on to the main sequence in much the same way that its parent did, but it will tend, on average, to be smaller and hotter. When it moves off the main sequence in old age it will be less likely to explode and more likely to pass away quietly to become a white dwarf. The important thing about a second-generation star is, however, not that its lifestyle is liable to be radically different from a parent star, but the fact that in the process of condensing from dust containing many elements it may become not just a sun, but a *solar system,* complete with planets.

This account of the lives of the stars is deliberately vague. I have seen several accounts much more definite, and perhaps clearer, but they tend to be mutually contradictory, and I think it would be misleading to imply that we can construct a biography for a star with any degree of certainty. The information which we possess concerning the stars is really as slight as the information we can obtain concerning subatomic particles, although it is evidence of a very different kind. The evidence on which we base our understanding of the stars is, in a sense, direct – what we know is carried to us by photons emitted by the stars themselves, and though our analysis of the information carried by those photons is as interpretative and as dependent on our own standpoint as our analysis of the information carried by any other photons, it provides us with clear, fundamental data. The indirect evidence we possess concerning events in the microcosm is somewhat less satisfactory – but on the other hand, reliance on indirect evidence goes together with the ability to experiment: we can endeavour to get the evidence we *want* rather than having to put up with what comes naturally.

For these reasons the limitations on what we can know about

the macrocosm are rather different from the limitations determining what we can know about the microcosm. In studying the macrocosm, we are immune from the uncertainty principle, for nothing we can possibly do can alter the properties of the things we observe. And yet we can still never know enough – what we know may be nothing but the truth, but we can never be sure that it is the whole truth, and if we are willing to doubt the assumptions attendant upon our egocentricity, then we may be very certain that it is not.

In passing on, therefore, from this brief account of the atoms of the macrocosm to a discussion of modern ideas about the universe as a whole, I think it is important to remember that the data in which these ideas are founded are incomplete and probably inadequate. It may well be just about all the data we are ever going to have access to, but that does not entitle us to ignore the existence of the rest or discount it as irrelevant. The nineteenth-century scientists cast in the Haeckelian mould believed that what they knew had to be all that there was to know. So far as they were concerned there was, as Berthelot said, 'no more mystery about the universe'. In the twentieth century, however, we have rediscovered mysteries, as the following pages will show. And perhaps – only perhaps – we should not go into this account assuming that it is the time-worn riddle of the universe all over again, searching as we go for one more tricky answer. Perhaps we should reflect instead on those facets of the nature of the universe of which we have no knowledge at all.

The idea of the expanding universe began, as so many discoveries in twentieth-century physics have begun, with a mathematician working out the implications of another man's equations. In this case, it was Willem de Sitter, who investigated the model of the universe proposed by Einstein in his general theory of relativity.

Einstein had represented the universe as a four-dimensional 'hypersphere' and co-opted Riemann's geometry to describe it. According to this model a ray of light did not travel in a straight line but along the arc of a circle. De Sitter argued, in 1917, that the hyperspherical model was not the only one which could be derived from Einstein's preconceived notions about what the universe ought to be like, or which could be used to explain the facts which relativity accounted for. He argued that the universe might actually be in a state of constant evolution, geometrically speaking, and that it might actually be intermediate between

Riemannian space and Euclidean, evolving from the former towards the latter. On this model, the path followed by a ray of light would not be circular but would be an ever-expanding spiral. If this were so then the distance between all celestial bodies would be gradually increasing.

De Sitter's new model came at the end of a historical period in which the philosophy of change had triumphed over the idea of static existence in many fields: the twin concepts of evolution and social progress had emerged as very powerful ideas within human attitudes to the world. For that reason, perhaps, de Sitter's suggestion was a natural one, and to some extent an attractive one. It was suggested by other physicists that the Einsteinian model might be unstable and must inevitably begin to change as de Sitter had specified.

Meanwhile, some rather curious observations were being made by optical astronomers. The measurement of Doppler shifts which became popular in 1912 led Vesto Slipher to begin calculating the relative radial velocities of certain nebulae. The Andromeda nebula – the first of those he measured, in that year – showed a small violet-shift. The Andromeda nebula was, apparently, approaching Earth at 125 miles per second.

Over the next five years, Slipher measured the radial velocities of fifteen spiral nebulae, and his figures were beginning to show a distinct bias: of the fifteen only two possessed a violet-shift. All the rest had red-shifts, and were thus moving away from Earth. After 1917 – the year that de Sitter published his speculations – Slipher found no more violet-shifts at all.

It was not until the mid-1920s that the calculation of distances via the study of Cepheid variables revealed that virtually all the nebulae whose red-shifts Slipher had determined were not local, but were, in fact, other galaxies. The two with violet-shifts, on the other hand, were members of our own galactic cluster. Slipher's work was taken up by others, and as more and more galaxies were investigated greater and greater red-shifts were measured. In the meantime, Hubble's calculations of the distances of the various galaxies allowed a comparison to be made between two sets of data. A strong correlation was found between distance and velocity of recession: the further away a galaxy is from us, the faster it is moving away. This became known as 'Hubble's Law'.

The first impression created by this revelation is that our galaxy is somehow special – cursed with a kind of galactic BO – but this is not so. *All* distances between *all* galaxies are increasing. It is rather as if they were all distributed over the surface of a balloon which is being slowly inflated – all of them are gradually growing apart from one another. This was exactly what had been suggested in de Sitter's model – that space possessed an inherent property of expansion. The situation seemed clear enough: de Sitter had predicted an expanding universe and optical astronomers had found one. The question was, however: had they found the *right* one? Mathematics suggested that, like Anderson finding the wrong meson, they had not. It must be remembered that de Sitter's prediction dealt not with space *per se* but with Einsteinian space-time, and that the 'de Sitter effect' was as much a slowing down of time as a stretching of space. This does not affect the measurement of the red-shift, but it would affect the measurement of distance by correlation of the periods of the Cepheid variables with their luminosity. When this was taken into account it seemed that the de Sitter effect was not nearly great enough to account for the observed red-shifts. Observations suggested that the galaxies were doubling their distances every 1,300 million years – not a vast interval on the macrocosmic time-scale (the Earth may be four times as old as that).

Catastrophists were naturally drawn to the idea that the universe was not simply expanding but exploding, and by extrapolating back they came to the conclusion that the origin of the universe was a cataclysmic explosion of a 'cosmic egg'. Others, convinced that de Sitter was right in principle even if the equations did not quite work, thought that the expansion might be 'built-in' to the universe and that new matter might be created to fill in the gaps between galaxies as they developed, thus preserving a 'steady-state'. As always, there was an obvious compromise between catastrophism and uniformitarianism, and the idea of an 'oscillating universe' which expands and contracts periodically *ad infinitum* was proposed (with no other reason than to satisfy aesthetic preconceptions).

The question was inevitably raised of whether the increasingly large red-shifts measured throughout the 1930s and 1940s were really attributable to velocities of recession at all. Was the expanding universe model really competent? Hubble's Cepheid yardstick was really only applicable to a small number of near galaxies, and where it no longer applied the only estimate of distance (apart from Hubble's Law itself) was the rather unsatisfactory method of comparing exceptionally large stars at the limits of luminosity. In fact, once Hubble's Law became established it quickly took over as the principle instrument for estimating the distance of remote objects. How reliable were the methods by which Hubble had determined the Law in the first place? Strictly speaking, not very. Studies of Cepheids had not shown them to be the most reliable of yardsticks individually, although on average the correlation between luminosity and period had worked out well – but one re-classification had had to be made when the Cepheids were sorted out into two types obeying slightly different correlation curves. Scientists are always reluctant to abandon methods that have led them to significant discoveries and elaborate theories, even when the methods come to seem unreliable – the discoveries and the theories become, by a circular argument, supporters of the methods. But can there possibly be an alternative explanation of the ever-increasing red-shifts?

It must be remembered that despite its name 'red-shift' has nothing to do with the colour of light coming from distant galaxies – it is a measure of the extent to which spectral lines have been shifted towards the low-energy end of the spectrum. If we are to assume that this is not accomplished by the 'stretching' of the waves emitted by incandescent hydrogen, then we must either account for the hydrogen emitting photon of the 'wrong' energy, or we must account for the photon's losing energy *en route* between the galaxies. If we are to assume that a hydrogen atom in a distant galaxy does not behave in the same fashion as a hydrogen atom on Earth, then we are saying that the universal constants science has discovered are not universal constants at all, but variables – and that knocks the bottom brick out of the whole edifice of scientific thought: the basic assumption that beneath the complexity of singular events there are orderly, unchanging principles. The second alternative – that the photons lose energy *en route*, also challenges the fundamentals of physics. If the photons lose energy, where does it go? It must go somewhere, if we are to retain the law of conservation of energy, and we have no concept of any process by which such a loss of energy might take place.

It is because of the extreme difficulty of framing an alternative hypothesis that the idea of the expanding universe holds so firm, and that Hubble's Law remains one of its cornerstones. It seems a reasonable enough idea – but it has led us into some very strange imaginative territory, and faced us with one or two apparent paradoxes. The most intriguing of these conceptual difficulties is the mystery of the quasars.

In the late 1920s the Bell Telephone Laboratories commissioned Karl Jansky to locate the source of static bedevilling their transoceanic radio-telephone operations, with a view to eliminating it. Jansky finally discovered that the main source of the 'noise' was a patch of sky in the constellation Sagittarius. He concluded that the interference could not be eliminated as it appeared to be coming from the centre of the galaxy. He and his employers then lost all interest in the phenomenon.

The idea of using these cosmic radio-signals as astronomical data was taken up by Grote Reber, who built himself a thirty-foot antenna in his back yard and set about mapping the chief radio sources in the sky — thus proving that even in the twentieth century the day of the lone amateur in science is not yet done.

It was after the war that radio-astronomy caught on in a big way. In principle, it was exactly the same as optical astronomy: measurement and interpretation of photons to glean data concerning their sources, and it opened up the possibility of using the entire electromagnetic spectrum in astronomy instead of just the limited optical range. (Thus, in the wake of radio-astronomy, we now have X-ray astronomy, infra-red astronomy and even gamma-ray astronomy.) The equipment used for picking up radio signals from space is rather different from that used in optical astronomy and involves special difficulties — particularly the accurate location of sources. Dipole antennas half a mile long or dishes 250 feet across are needed for tracking spacecraft or pinpointing distant sources. Exact location requires 'binary vision' — pairs of dishes set miles apart compare signals and focus them.

Radio-astronomy gives a very different picture of the sky from optical astronomy. Individual stars are poor radio-emitters, although the sun and some of the planets emit enough to be detectable (and these radio-emissions have been very influential in forcing us to re-appraise our ideas about what some of the planets are like). But it is with more distant regions of space that radio-astronomy really comes into its own, for photons at such frequencies go clean through the big dust-clouds which absorb visible light from far sources. The galactic nucleus, invisible to optical telescopes, is quite easily detectable by radio-telescopes. The most interesting sources of all, however, were the 'radio-stars' — apparent point-sources which could not at first be correlated with visible objects.

When Baade associated one such source with an optical object which appeared to be two galaxies in collision, the imagination of radio-astronomers was set alight. What wonders might not the other radio-stars reveal?

Many of the radio-stars did, in fact, turn out to be very distant radio-*galaxies*, including many anomalous ones. Galaxies in collision did not seem to be common but one of the radio-galaxies (M82) seemed to be in the process of undergoing a cataclysmic explosion – a kind of galactic supernova (and a sight to gladden the eye of any hardened catastrophist).

As it became possible to determine more and more accurately the position of radio-sources more anomalies began to arise. Some of the apparent point-sources, it transpired, really were very small – much smaller than galaxies. Hazard tracked one of these sources, 3C273 (i.e. object no. 273 in the third Cambridge catalogue) while it was occulted by the moon. When the moon's rim cut off the signals it did so momentarily, and Hazard was able to show that the radio-source corresponded exactly with a star-like object – certainly not a galaxy. Optical astronomer Maarten Schmidt then inspected the spectrum of 3C273 and found it extremely peculiar. At first it defied explanation, but was later shown to exhibit a red-shift that was quite colossal – characteristic of the furthest and faintest galaxies. It was, in any case, obviously no ordinary star, and it was dubbed a 'quasi-stellar object', which later became shortened through jargonisation to 'quasar'.

Other similar objects were soon investigated, and the red-shift of 3C273 was soon revealed to be no fluke – in fact, it almost paled into insignificance. Within ten years of the first quasar identification (in 1963) hundreds more were detected. Two of them – 0H471 and 0Q172 – had red-shifts which translated into velocities of recession of $0.9c$ – and, if Hubble's Law was applicable, must therefore be the most remote objects known. And yet they were point-sources of light, not diffuse sources, like galaxies – and they emitted enough energy to be visible in the optical range.

If the quasars are really as far away as Hubble's Law suggests, then their luminosity is inexplicable – and, indeed, almost incredible. On the other hand, if they are local objects, what is the reason for their extreme red-shifts? Why do they appear to be participating in the general expansion of the universe?

The quasars present a serious challenge to Hubble's Law, and there have been attempts recently to expose it as a jumped conclusion. Halton Arp has attempted to show that distance and red-shifts do not always tally with respect to the nearer galaxies whose distance can, in fact, be estimated with some degree of certainty – he has argued that in one case Hubble's Law may be wrong by a factor of eight. Other astronomers have pointed out that galaxies which are so very close to one another in the sky as to seem associated often have widely discrepant red-shifts, suggesting that they are strung out across half the universe. Geoffrey Burbidge has argued that some quasars are suspiciously close to galaxies which show much lower red-shifts than they do, and there are even quasars which seem to be 'surrounded' by the members of a low-shift galactic cluster.

Catastrophist theories suggest that quasars may be imploding galaxies, and seek to explain the red-shift with reference to the velocity of collapse rather than the velocity of galactic expansion. But the quasars all seem to be star-sized sources, and their radiation is often slightly unsteady, suggesting that they are objects rather than aggregations. Also, there are far too many of them to be explained by such a short-lived phenomenon.

In 1965 Sandage discovered a class of bluish stars which showed massive red-shifts but which were not radio-emitters. These 'blue stellar objects' may be more numerous than quasars – and may, indeed, be quasars at a different stage in their own evolution (i.e. 'aged' quasars). At present, these are things which are simply not accountable by our theories. We must somehow expand or change our ideas about the universe to accommodate them. But exactly *how* do we change our ideas? What do we add or replace? We have no evidence to help us except the enigma itself.

There are other implications to be drawn from Hubble's Law and the challenge of the quasars. If, as the galaxies draw further and further apart, their relative velocities grow even faster, then ultimately they must – relative to Earth – approach the velocity of light. Some known quasars, in fact, appear to be doing just that: their velocities are in excess of $0.9c$ and they are presumably still accelerating. As they do so they must, relative to Earth, become foreshortened and slowed in time. At the edges of *our* observable universe (they will, of course, still be at the 'centre' of their own observable universes) they will become wafer-thin relative to us, packed ever-more densely into a layer of cosmic wallpaper which, though infinite in extent, is also infinitely compressed. There is no distance beyond the distance at which the velocity of recession tends towards the limiting c, but there is still an infinite universe there.

But as the galaxies continue to draw apart, there must, in the fullness of time, come a stage where our galaxy is left alone in its observable universe, with nothing but a great gulf extending to the perceived horizons. There must, that is, *if* the big bang theory is correct and not the steady-state theory proposed by Hoyle, Bondi and Gold.

It should be noted that the assumption of the steady-state theorists is precisely the same as that adopted by Einstein in deriving the theory of relativity: that the universe ought to look the same from any standpoint in space-time. This is known as 'the perfect cosmological principle' and is, of course, purely aesthetic. If the steady-state universe is incompetent (and it violates that other beautiful cosmological principle, the law of conservation of mass-energy) then there must surely be some reservoir of doubt about the competence of relativity.

And it seems, in fact, that the steady-state model is incompetent. There is really only one way for us to decide between the big bang theory and the steady-state theory, and that is to observe the universe over a long (by macrocosmic standards) period of time and see whether it does, in fact, show signs of directional change. If such directional change could be observed, then the steady-state model is disqualified.

Thanks to radio-astronomy, we can actually do this, for time is only an aspect of space-time, and thus when we look into the vast depths of space (vast, that is, by macrocosmic standards) we are actually looking into a vast depth of time. The photons which come to us from local space are recent photons – but those which come to us from the edge of our observable universe were emitted billions of years ago. In looking into the very furthest reaches of space, therefore, we are looking at a much 'younger' universe. Is it any different from our own local space?

The answer seems to be *yes*. If the red-shifts of the quasars can be translated into distance *via* Hubble's Law, then the edge of the universe is rich in such objects relative to local space. And this argument also offers us a possible conceptual escape from the paradoxical implications of the quasars: if they do belong to a 'young universe' and no longer exist in our own, more mature one, then they might well be objects concerning whose nature we can find no hint in our observations of the kind of thing which happens in our own galaxy. This idea does not help us to explain the quasars but it helps us to see why we *cannot* explain them with our present understanding.

There is a certain circularity about this whole argument. If the quasars are *not* at the distance suggested by their red-shifts, then our grounds for throwing out the steady-state model of the universe are rather shaky. But there remains one other piece of corroborative evidence. Theorists building mathematical models of the big bang suggested that the energy of the big bang itself might still be perceptible today. Such tremendous energies as might be required to explode a cosmic egg would have resulted in the radiation of very hard X-rays, which would be red-shifted all the way into the microwave region of the electromagnetic spectrum as they come to us today from the furthest reaches of the universe. In 1965 the Bell Telephone Laboratories announced that after all other accountable radio sources were allowed for, there was, in fact, such a 'background' radiation: the heat of the explosion, several billion years later. The observation is not conclusive − for one thing, the microwave frequencies do not quite tally with the frequencies predicted by the theorists − but it may well be indicative. The perfect cosmological principle may be a write-off, and the applicability of aesthetics to cosmology is thus challenged in much the same way that its applicability to atomic theory was challenged by the fall of parity.

The discovery of quasars and the subsequent disturbance of our understanding of the macrocosm is analogous to the discovery of the host of subatomic particles which followed the development of new observational methods after the war. And, just as the pure theorists in the field of atomic physics kept pace with the observers by playing with even more ambitious imaginative concepts (quarks, gravitons, etc.), so the pure theorists in the field of cosmology have not allowed the quasars to snatch all the imaginative glory. Indeed, their newly fashionable, purely hypothetical addition to the conceptual universe has probably seized the popular imagination far more powerfully than anything in microcosmic physics. I refer, of course, to 'black holes'.

Actually, the basic concept of the black hole is older than the de Sitter model of the expanding universe. Karl Schwarzschild, in 1916, examined by thought experiment some of the consequences of Einstein's supposition that light was subject to gravity. He wondered how far the mass of the sun would have to be compressed before it was so dense that light could no longer escape from its surface. He calculated that this limiting radius for a body of the sun's mass was about three kilometres. The sphere described by this radius (known, naturally, as the Schwarzschild radius) would then constitute an 'event boundary', because whatever went on within it, nothing could happen which could in any way influence or be observed by anything outside it. Only by the gravitational field extending out from the event boundary could the body be perceived at all, and matter falling through that field would be swallowed up by the event boundary – essentially abstracted from the observable universe. It would literally have 'fallen through a hole in space'. The greater the mass within the hole, the greater the Schwarzschild radius would be, and therefore the greater the surface area of the event boundary (i.e. the mouth of the hole) – and its mass-swallowing potential. A black hole big enough might therefore swallow whole stars, and – growing steadily more ravenous as its mass and mouth increased – whole star clusters.

As a thought experiment, Schwarzschild's speculations did not attract much attention. But astronomical discoveries in the post-war years began to redirect the imaginative interest of cosmologists towards the concept.

Firstly, the white dwarf stars were identified as degenerate matter, and the ultimate density of normal matter was surpassed: atoms could be collapsed. The possibility of a further stage of degeneracy – to the point where the atoms were completely crushed and became neutronium – became an open question. It was not too long before this concept had to be called up to explain certain strange entities named pulsars: sources of radio-waves which varied sharply and within a very fast period. The short period of the signals suggested that the objects in question were rotating very rapidly – and in order to do so they would have to be very small indeed. One pulsar – at the heart of the Crab nebula, wreck of the supernova of 1054 – was finally identified with a visible object. It was not a white dwarf, and it was concluded that it must be a neutron star. The pulsating property was explained as a 'searchlight effect' caused by a powerful magnetic field.

Once the existence of neutron stars was established as a strong possibility, then the existence of black holes began to seem quite likely. A neutron star massive enough would inevitably collapse within the Schwarzschild radius – and so some supernovas may well leave a black hole behind them.

The black hole is a most useful concept to have been added to cosmological theory, because it has great potential as a scapegoat for unexplained phenomena. It is rather akin to Pauli's neutrino in that it is virtually undetectable, so that one can put a black hole exactly where one wants in order to do exactly what is needed to explain an anomaly. For instance, observations of Pluto – which was initially invented in order to explain perturbations in the orbits of Uranus and Neptune long before being seen and identified by Tombaugh in 1930 – reveal that it is very small; so small, in fact, that if it is really guilty of the distortion attributed to it it must be far denser than any ordinary matter. This difficulty is usually shunted aside in tables giving the vital statistics of the solar system – Pluto only gets question marks for size and mass. Some astronomers were willing to hypothesise a tenth planet which was *really* causing the trouble, others were simply willing to let the matter rest. But with the aid of a black hole the problem can be made to disappear almost magically: obviously, there is a black hole at the centre of Pluto, and that explains its great mass.

The black hole in question would have to be very tiny – but a very tiny black hole would still be very massive. And the notion of a solid body with a black hole at its centre is certainly not incompetent: a black hole and a massive body meeting in space would literally fall into one another. If the black hole were very small the rate at which it would consume the solid body might be very slow – a few atoms a year, increasing very gradually. In the meantime, the hole would settle at the centre of gravity of the body.

It has also been suggested that the Earth met a tiny black hole in June 1908 when a tremendous explosion is said to have occurred at Tunguska, Siberia. No crater was formed, as would have been expected from any physical impact. But if the Earth met a tiny black hole (about one millionth of a centimetre in diameter, having the mass of a small asteroid) perhaps the event can be explained – the black hole must, though, be credited with enough velocity to take it clean through the Earth and out into space again.

Such tiny black holes are, it seems, very versatile things. It has been suggested that the matter in the solar system might have condensed around tiny black holes (thus explaining the structure of the solar system as a product of its initial black hole distribution) and that all the planets have black holes in their cores, consuming them at a very slow rate. The centre of the galaxy, about which it revolves, might be a black hole – and so might the rotational centre of the observable universe. All this is, of course, pure speculation – but it is interesting to see how a good tricky answer to the riddle of the universe circulates almost as fast as a funny joke. There is a big theoretical 'market' for black holes if we can ever find one. The big demand is for tiny ones rather than the very massive ones formed as the relics of supernovae and optimists are already trying to build mathematical models which permit billions of tiny black holes to be formed or liberated by the explosion of the cosmic egg.

The properties of black holes are both fascinating and mind-stretching, and it is most interesting to conduct the thought experiment of falling into one.

A distant observer watching someone falling into a black hole would see him fall ever more slowly as his velocity seemed to approach *c*. The observer would never see him actually hit the event boundary because the light-rays carrying information to him would always come from just outside that boundary – grossly red-shifted and carrying images much extended in time. From the viewpoint of a distant observer, therefore, the fall would never end (in much the same way that from the viewpoint of an observer on Earth a spaceship accelerating away from Earth could never be seen to reach the velocity of light).

The view of the man who was actually falling would, however, be very different. To him, light would appear to be behaving perfectly normally (in accordance with Einstein's principle) and he would complete his fall quite quickly – the 'same' span of time is finite for him but infinite for the observer – and cross the event boundary. What happens then? What is going on inside the event boundary?

Gravity is a 'curve' in space, and when a gravitational field becomes sufficiently intense (when a mass is compressed within the Schwarzschild radius) the curvature becomes circular and 'sealed'. This state of closed curvature is called *singularity*. What has happened within a black hole is that the space within the event boundary has become a universe in its own right – finite in volume so far as our universe is concerned, but *in*finite with respect to its own internal frame of reference. Earlier, I pointed out that our observable universe is finite in extent but infinite in volume because as the receding galaxies approach light-speed relative to ours they become ultimately thin. This 'cosmic wallpaper' might be regarded as the 'inside' of an event boundary – and, by a striking coincidence, if we take the radius of the observable universe as calculated by Hubble's Law and calculate the mass required to be concentrated within the radius in order to qualify it as a black hole according to the Schwarzschild equation, the result is about the same as the estimated mass of the universe calculated from the density of intergalactic matter.

The observer falling into the black hole would be literally entering a different universe. He could only do so, of course, as a hypothetical observer in a thought experiment, because if he had any real substance in him it would be crushed into neutronium and 'translated' into an entirely new frame of reference.

In the concept of the black hole we have, once again, the worlds-within-worlds idea which first emerged from the solar system/Bohr atom analogy – but this time it is in a form far more appropriate to twentieth-century physics. There is an imaginative glamour about the idea of black holes within black holes within black holes, especially in the idea that one cannot get from one to another by mere vulgar shrinking, but only by complete translation into a whole new kind of existence 'far beyond the limits of uncertainty'.

The best chance we have at the moment of actually locating a black hole is by investigating the dark companions of certain binary star systems. Such a system might reveal the presence of a black hole through the anomalous behaviour of the visible component in the hole's gravitational field. Theorists, however, have not waited for observation to catch up with the standard spherical black hole before going on to contemplate the possible peculiarities of non-spherical ones. There is the particularly interesting case of a ring-shaped hole formed by the collapse of a large mass rotating very rapidly. In this case, in the plane of the ring there might be no event boundary, and so the singularity within the hole would – from certain viewpoints – be unprotected by its curtain of 'frozen time'; it would, in fact, be a 'naked singularity'. If such a thing does exist, then we would have a window into another universe – another frame of reference.

The black hole and the uncertainty principle are, in a sense, the 'brackets' of our conceptual universe, just as the wavelengths of red and violet light are the brackets of our sensory universe. But what kind of limits are they? Do they really set limits to what we *can* know, or is there some intellectual means by which, just as we can translate the invisible into the visible, we can translate the inconceivable into the conceivable?

We are at least willing to try. There have been attempts to construct, hypothetically, other universes populated by entities which could hardly exist in ours. Perhaps the most interesting such hypothetical construct is the tachyonic universe, which mirrors Einstein's in that the velocity of light becomes a *lower* limit, never to be reached by deceleration. A tachyon in our own universe would travel 'backwards' in space-time, reaching its destination before it set out. Anti-matter, too, has been 'explained' as matter orientated backwards in time: thus, a positron, which is like an electron except that it is deflected the wrong way by a magnetic field, could be visualised as an electron being deflected the right way, but travelling backwards in time. These ideas are the products of mathematical aesthetics, pure and simple, and they testify to the fact that mathematical aesthetics transcends what we would normally describe as 'the possible' and may allow us to describe concepts in, and bring some order to, the realms of the *impossible*. The limits of the observable universe are not necessarily the limits of the imaginable universe.

But this is, perhaps, to get too far ahead of ourselves. On the other side of the coin is the more disturbing question of whether we can actually aspire to understand the observable universe. Have we the mental equipment which will allow us to come to terms with *non*-hypothetical entities? Can we accommodate the quasars and their kin within our models of reality?

It may well be that the basic assumptions on which all our scientific knowledge is built have not yet been modified enough by twentieth-century theory. Newton's absolutes have gone, but Einstein's cosmological principle remains, and whatever the victories of understanding to which that principle has carried us, it remains an assumption.

The first accounts which men constructed to explain the world in which they found themselves were not scientific at all, but mythical. Every race, every culture, evolved by inspiration and intuition its own accounts of cosmogenesis and its own scheme of divine laws. Even some of the early Greek scientists welcomed alternative explanations for natural phenomena – they did not see why there had to be one set of explanations and one only. Science is a form of monotheism with a mathematical godhead. But science – the assumption that there is one set of systematic ordering principles: rational, unchanging and universally applicable – may be just one more myth, whose exponents are claiming it as absolute truth in exactly the same way that the Christian Church proclaimed its beliefs in endeavouring to convert the heathens and the pagans of the world.

It seems to me that the deeper implications of the de Sitter theory of the expanding universe are perhaps insufficiently appreciated. He suggested that the properties of space-time itself are subject to change. It remains possible that the anomalies we see in looking into the furthest reaches of the universe, which prove so difficult to interpret, may be the result of some such process of change.

De Sitter himself wrote the following:

After all, the 'universe' is an hypothesis, like the atom, and must be allowed the freedom to have properties and to do things which would be contradictory and impossible for a fine material structure .... The conclusions derived about the expanding universe depend on the assumed homogeneity and isotropy – i.e. on the hypothesis that the observed finite material density and expansion of our neighbourhood are not local phenomena but properties of the 'universe'. It is not inconceivable that this hypothesis may at some future stage of the development of science have to be given up, or modified, or at least differently interpreted.

Like twentieth-century space and time, twentieth-century scientific knowledge is not absolute, but relative. We must not forget – as some scientists are occasionally prone to do – that everything we know is built upon assumption, and is thoroughly intermingled with what we think and what we believe. Modern scientific knowledge is an impressive edifice – but even the most perfect edifice may collapse completely if it proves to be founded in quick-sand, or if the ideative cement holding together its factual bricks crumbles in the heat.

Part Two
The Tree of Life

For having, in the natural history of this earth, seen a succession of worlds, we may from this conclude that there is a system in nature; in like manner as, from seeing revolutions of the planets, it is concluded that there is a system by which they are intended to continue those revolutions. But if the succession of worlds is established in the system of nature, it is in vain to look for anything higher in the origin of the earth. The result, therefore, of this physical inquiry is, that we find no vestige of a beginning — no prospect of an end.

James Hutton, *Theory of the Earth*

What a book a devil's chaplain might write on the clumsy, wasteful, blundering, low, and horribly cruel works of nature!

Charles Darwin

5   The Discovery of Change

There is an imaginative division between the physical sciences and the biological sciences. The pursuit of understanding in the realms of physical science is essentially an intellectual one, in which man participates as mind, standing outside the categories of events which he tries to systematise and rationalise. But the biological sciences – the sciences of life – put rather greater demands on their theorists. Any man who attempts to put the world of living things into a new imaginative context, and to change our attitude to them, must, by implication, make a new image of man himself. In the biological sciences, man is not only the observer, but a member of the class of entities which is being observed. We cannot describe life without describing man, and the way that the scientist conceives of life cannot help but provide his concept of man.

Beliefs, whether they are religious or scientific, are like mirrors in that they contain images rather than real things. Biological science aspires to be the plane mirror that reflects 'accurately', without distortion, rather than the curved mirror which magnifies or the magic mirror which tells its owner who is the most beautiful in the land. During the last century the image of man reflected in the mirror of the biological sciences has changed drastically, in the wake of Darwin and Freud.

Before Darwin, there were two principal images of man – neither of which was determined by scientific beliefs. There was the Christian view of man – a being made up of an immortal soul wrapped in a fleshly envelope. There was also the image which had emerged during the Enlightenment of man the sociable and perfectible being, motivated by rational self-interest. Darwin undermined the first opinion, Freud the second, and they colla-borated in producing a new image of man: 'natural man', the product of evolution-by-chance, his behaviour largely at the mercy of his evolutionary legacy and the unconscious mind.

The discovery of the new image of man was intimately bound up with a host of new discoveries about the nature of the world in which we live – the thin layer of activity at the interface of the solid, liquid and gaseous components of the Earth. The most important aspect of this revolution in attitude was the replacement of static concepts of the living world with dynamic ones: the discovery of the nature of change.

In the Middle Ages men believed that the heavens were eternal and unchanging, the finished product of divine design. On the Earth itself they recognised processes of change only within a grand design in which *plus ça change, plus c'est la même chose.*

One generation passeth away, and another generation cometh: but the earth abideth forever.

The sun also ariseth, and the sun goeth down, and hasteth to his place where he arose.

The wind goeth toward the south, and turneth about unto the north; it whirleth about continually, and the wind returneth again according to his circuits.

All the rivers run into the sea; yet the sea is not full; unto the place from whence the rivers come, thither they return again …

The thing that hath been, it is that which shall be; and that which is done is that which shall be done; and there is no new thing under the sun.

So says Ecclesiastes, the preacher of the Old Testament, and in such a context man placed the cycle of the seasons, and history, and his own mortality. So strong was this conviction that change was nothing but cyclic recurrence, that generations of scientists took for granted the fact that all natural motions must be circular. Science took two thousand years after Aristotle to get to grips with the mechanics of motion because the idea of *linear* motion contained within Newton's first law was so alien.

The Copernican revolution which culminated in Newton's subjection of the heavens to the law and order of natural process instead of divine ordinance was, in a sense, only half a revolution in human thought. In the same year (1543) that Copernicus's heliocentric theory made its first tentative appearance in print Vesalius published his anatomical account of the human body. The first microscopes and the first telescopes appeared at the turn of the same century – within a decade of 1600. But the translocation of the living world into a 'natural' context alongside the heavenly one was not achieved in the seventeenth century. In fact, one important factor in the victory of Newtonian synthesis over dogmatic Aristotelianism was a compromise between Church and science by which the heavens were ceded to science while the Church retained its intellectual monopoly of the human world.

Descartes divided the conceptual universe into two: the physical universe and the spiritual. The world of matter, according to Descartes, was the appropriate realm of scientific philosophy, while the spiritual and moral world was properly subject to religious thought. The scientist was entitled to make authoritative statements about *things*, but it was for the Church to discourse on the matters of the human soul.

This 'Cartesian partition' smoothed the way considerably for a new harmony between churchmen and physical scientists. What it failed to do, however, was make things any easier for biological scientists. God the prime mover and engineer of the heavens had been made redundant, but God the creator and architect of life was still very much in business. Physicists were no longer in danger of stepping on the toes of the Church, but biologists were all the more likely to incur displeasure because they were trespassing on the prerogatives of the Almighty instead of sticking to the inquiries that were properly theirs. And so science 'conquered' the heavens two centuries before it won the Earth.

In the seventeenth century it was taken for granted that all changes which had occurred on the face of the Earth had been the handiwork of God: specifically, the old world had been destroyed by the Biblical Deluge. Even then, however, men like Thomas Burnet were willing to countenance the notion that other changes had come about through the action of 'subterraneous fires', earthquakes, wind and rain. The logic behind Burnet's belief was the idea that it was inconceivable that God should have created such an untidy planet – it must have deteriorated since the Creation.

Burnet's contemporary James Ussher, Archbishop of Armagh, undertook to investigate the question of the age of the Earth by closely studying the chronology of the Old Testament, and came up with the answer which made him famous: that the Creation took place in the year 4004 BC. Already, however, there were other voices to be heard – notable among them that of Robert Hooke – who suggested that the changes which had taken place on the Earth's surface must have taken quite a long time. Hooke was convinced that fossils – the collecting of which was becoming something of a popular hobby – were the remains of real plants and animals, many of which were marine species. In his view the Deluge did not suffice to explain why certain areas which were now land seemed once to have been under the sea – nor had it been of sufficient duration to bring about the profound changes attributed to it.

Perhaps it was as early as this that opinion as to how the Earth's surface had come to assume its present aspect began to polarise between catastrophist theories and uniformitarian ones, but certainly the catastrophists were first on the scene, and the Deluge was clearly the popular favourite.

It was the Comte de Buffon, whose *Natural History* began publication in 1749, who first declared that scriptural accounts of the history of the Earth should be ignored, and that the aspect of the surface of the Earth must be explained in terms of the effects of natural processes operating on an everyday scale over a long period of time. Like Galileo before him, Buffon was called upon to recant his heretical ideas, and in the fourth volume of his work (1753) he retracted the opinions set forth in the first. Like Galileo, however, he had more respect for the implications of his own observations than the dogmas of the Church, and twenty-five years later, he published *Epochs of Nature*, in which his uniformitarianism was declared for a second time. In this book Buffon identified six Earthly Epochs which, though of indeterminate length, covered a great many thousands of years, and multiplied Ussher's calculated age of the Earth nearly a hundredfold. In order to try to appease the clergy, he suggested that the 'six days' in which God had created the Earth were not days in the literal sense but periods of indefinite length. The clergy were not impressed.

Even Buffon's time-scale was not really adequate to enable natural philosophers to fit the data they were trying to deal with into a proper context. In 1795, however, there appeared a new uniformitarian theory, proposed by James Hutton. It was Hutton who attempted to find a 'system in nature' akin to the system which Newton had found to regulate the motions of the heavenly spheres. And he found one – a tripartite process causing slow but constant change in the surface of the Earth. First, he decided, regular strata of rock were formed by gradual deposition in the ocean. These strata were then consolidated and elevated by subterranean pressure. In the meantime, wind, water and organic decay weathered the face of the Earth and washed the land slowly into the sea. This picture of the mechanics of geological change was an important contribution to natural philosophy in its own right, but even more important was the new perspective at which he arrived in the conclusion of his inquiry, which was simply that, given the system, the question of putting an age to the Earth became redundant. 'We find', he wrote, 'no vestige of a beginning, no prospect of an end.'

It was Hutton's views which put geological thinking on to the proper time-scale, and established an imaginative context for the idea of evolution. Before Hutton published his *Theory of the Earth* there had been considerable debate regarding extinct species. The fossils known to Hooke, even when accepted as real organic relics, had not been seen as anything out of the ordinary. It was not until 1761 and the discovery of the remains of the mammoth – and, subsequently, the mastodon – that it became obvious that not all the creatures which had roamed the ancient Earth were still around in the eighteenth century. If species died, then the pattern of life on Earth was no more immutable than the pattern of stars in the sky. And once it was established that change *had* happened, it became an open question as to whether change was still going on.

The theory of successive creations was popularised by Cuvier in the early nineteenth century to explain the discrepancies between fossil fauna and the present population of the world. These successive creations, he argued, were each destroyed by catastrophe of one kind or another, and their distinctness was obvious in the clearness of the rock strata. His contemporary Lamarck, however, was a uniformitarian and quarrelled with the evocation of catastrophism to account for the fossil record. He and others, including Erasmus Darwin, saw in the fossils a *continuum* of living forms: a pattern of developmental progress. According to them, the changes which had occurred on the surface of the world, in a period of time to which no one could set a limit, were sequential; the many forms of life had *evolved*.

It was Lamarck who first attempted to unite all living forms into a single 'family tree', originating from a single common ancestor. It was Lamarck, too, who first proposed that the process of change which had made the tree grow was concerned with the *adaptation* of animals and plants to new environments and new ways of life.

Lamarck placed man at the top of his tree of life – the ultimate product of progress. But that privileged position was still, in the eyes of the Church, an ultimate degradation. The battle to define the status of man in the universe of life was joined, and became very bitter, decades before the younger Darwin made his crucial contribution to the argument.

I have gone into this background in some detail because I think it is extremely important to realise what the intellectual climate into which Darwin delivered his *Origin of Species* was like. Only by identifying the precise nature of the ideative conflict in which the book became a weapon can we understand why the subsequent history of Darwinian theory in the twentieth century is so very peculiar.

The debate between Church and science over the idea of evolution was really a battle to decide whose prerogative it was to make statements about man. Descartes, back in the seventeenth century, had left man at the very interface of the partition he tried to draw between the areas of religious and scientific concern: the human body, he had said, was a thing, a machine – but inside it, directing it from its seat in the pineal body, was a *soul*. Because of the soul, religion claimed man as the proper subject of religious thought; while science was determined that there was nothing sacred about the body.

Lamarck had suggested that man was descended from the animals, and that the process of change by which he had done so was adaptive. What he had not been able to do was identify the mechanism of adaptation. He was in a position analogous to that of Kepler, who knew exactly what the planets did, but – lacking Newton's theory of gravity and laws of motion – could not quite identify the systematic principles which determined their behaviour.

The naturalist philosophers, involved in the battle with the Church, desperately needed a weapon to give them the upper hand: they wanted a systematic account of how evolution happened, and without one they had little chance of disposing of divine providence. Darwin provided that weapon in the *Origin of Species*, and it was gratefully taken up. It became *the* great revelation of nineteenth-century science, but it did so far more because it was *useful* – indeed, desperately necessary – to the rationalist philosophers than because it was totally competent or convincing.

Darwinism is now established. It has been vindicated by experiment and observation. Natural selection works as an agent

The Discovery of Change 111

of change. But this should not conceal the fact that the theory owes both its origin and its initial acceptance to aesthetic and philosophical considerations quite apart from its scientific validity. In an earlier chapter, I made a similar point with respect to Einstein's theory of relativity, but there is one important difference between Einstein's theory and Darwin's, and that is the intellectual context to which they belong. If Einstein had been wrong, and his predictions had not been confirmed, it would have mattered very little. His ideas were purely scientific, addressed to the scientific community, where some theories live, some die, and some change as a matter of course. But if Darwin had been wrong, it would have mattered greatly, for *his* theory was a flag which the rationalist philosophers were waving at the world: the symbol of their authority over the whole of truth. Darwin *had* to be right, in order to establish the priority of science in guiding human thought.

But Darwin was not proved right overnight. His theory, as it appeared in the *Origin of Species* in 1859, was lacking in two important respects: he had no notion of the hereditary mechanism by which characters were passed on from one generation to the next, and he knew of no source for the random variations on which his theory depended. Mendel's theory of genetics, which supplied the first deficit, was discovered soon after but ignored by the scientific community, and had to be rediscovered in 1900. The theory of mutation, which supplied the second deficit, did not receive any experimental support until 1920 or thereabouts. It took, in fact, more than half a century for Darwin's theory to become conceptually adequate. In the meantime, it was supported by faith: the supreme faith of the nineteenth-century rationalists that they were right and the Church was wrong.

That faith survives today and is still strong. The battle between Adam and evolution is not yet over, although the balance is very much in evolution's favour, and 'Thou shalt believe implicitly in Darwinism' is still one of the principal rationalist commandments. But if Darwinism has proved competent, what does it matter? Why make such a fuss because its support comes largely from faith rather than from reason?

The answer to those questions is that because of its support by faith Darwin's theory of evolution by natural selection has been allowed to survive in its original form, unchallenged and unrevised. The idea which we have of evolution is to a large extent the idea that Darwin had – and it is a very biased view.

Fundamentally, Darwin was right, but simply because he did not have access to the knowledge we now have, he put the emphasis on a particular aspect of natural selection: the aspect described by the phrases which became household words – the 'survival of the fittest' in 'the struggle for existence'. In the twentieth century, it was demonstrated that this kind of natural selection could and does happen – but we have also discovered the potential for other kinds of natural selection which may have equal bearing on our interpretation of the evolutionary process – alternatives which Darwin did not consider. And *because* Darwin did not consider them, they have been neglected and not properly integrated into the theory, as they should have been, and would have been if Darwinism had been promoted as a scientific theory rather than as philosophical propaganda.

In this section of the book my dominant concern will be to try to explain how the theory of natural selection must be re-interpreted in the light of twentieth-century discoveries in the biological sciences.

Darwin developed the theory of evolution by natural selection by drawing an analogy between factors acting in nature and the strategies employed by animal breeders to 'improve' domestic stocks and select a variety of specific characteristics. It was the origin of the theory in this analogy which led to its innate bias. Animal breeders select the best of their stock for breeding purposes and, in the case of food animals, slaughter the rest for eating, or – in species like pigeons and cats, bred for specific ornamental characteristics – dispose of the less desirable specimens altogether. Darwin, while he was ship's naturalist aboard the *Beagle*, had made a close study of the fauna of the Galapagos Islands and had been particularly impressed by the way in which different species of finches occurred on different islands, adapted to wholly different ways of life. He compared these different finches to the different breeds of pigeon artificially obtained by domestic selection.

His main difficulty was in discovering some agent in nature which could fulfil the role of the animal breeder, rejecting the 'worthless' stock and pairing the 'best', and here his imagination was stimulated by Malthus's *Essay on Population*, an economic tract written in 1798 which pointed out that productivity could only increase linearly while population tended to increase exponentially unless restrained by plague, famine, war or whatever. This gave Darwin the heart of his theory. Natural populations, he decided, always produced more progeny than can possibly survive to give birth to a new generation: if they did survive, there would be a 'population explosion' which would far outstrip the available food resources. The young, therefore, were born into a constant struggle for existence in which there was only room for the strong to survive. The weak had to perish at the hands of predators, parasites or by their own inability to compete for food. Those which survived this struggle were, by definition, the fittest *to* survive, and by this means the 'quality' of the population was maintained and constantly improved. New species arose when, through the inherent tendency which animals had to vary, potential for adaptation to a new way of life was discovered, enabling some individuals to evolve apart from the rest of their species in pursuit of a different direction of improvement.

As I have already pointed out, the theory was incomplete. Knowing nothing of the mechanism of heredity, Darwin could not explain how characters were transmitted from one generation to the next, how potential for variation was conserved, and how it was that variation occurred at all. Nevertheless, the theory was intellectually compelling, for several reasons – some of which had nothing at all to do with biology. Darwin's inspiration had come from an economist, and his concept of evolution bore a very strong resemblance to the theory of capitalist economics. Indeed, the phrase 'the survival of the fittest' was not originally Darwin's at all, but had been used by sociologist Herbert Spencer in a social and economic context. Darwinism was therefore accepted most gratefully by the capitalist social scientists because it sanctified their theories with a halo of 'natural law'. This 'social Darwinism' was taking Darwin's theory right out of context, and applying it in an area where it was totally incompetent, but this was nevertheless one of the primary reasons for the rapid acceptance of Darwinism in some quarters. It was much easier to accept such a theory when one could use it as a logic to explain things which were happening to ordinary people in everyday reality. It was through its social analogues that Darwinism most clearly made sense in the popular mind. (It was some time before the limits of that particular kind of common sense were exposed in the real world when Hitler, armed with precepts of social Darwinism, set out to improve the human race.)

Curiously enough, Marx and Engels, who were in the process

of formulating a theory which was to become the principal alternative to capitalism in the socio-economic world, also welcomed Darwinism. The analogy which they drew was, however, rather different – so far as they were concerned, in proving that evolution was a natural process Darwin had established the naturalness of *social* evolution, whose direction, they believed, was towards socialism. (The Russian Communist Party, in later years, was to find the logic of social Darwinism so repellent that they were moved officially to abandon scientific Darwinism and revert to a neo-Lamarckian doctrine. They kept, of course, the notion that evolution was a natural and inevitable process in society.)

Those who were most ready to accept Darwinism, then, did so for reasons other than the scientific evidence which Darwin had mustered. And, of course, those who were most ready to attack and reject it also did so for reasons which had nothing to do with its scientific competence (or otherwise). The confrontation between Church and science with respect to Darwinism has come to be symbolised by the British Association debate in which Thomas Henry Huxley ('Darwin's bulldog' – the theory's most vociferous propagandist) faced Archbishop ('Soapy Sam') Wilberforce – a travesty of a debate which degenerated into personal abuse, and in which reasoned argument was not employed by either side.

Darwin himself was never as sure of his theory as its most dedicated supporters. He, at least, was scientist enough to be aware of its flaws and to harbour such doubt as was reasonable about its competence – doubt which he retained to the grave, in harness with doubts concerning the wisdom of publishing his ideas and precipitating the storm. He did, in fact, delay publication for many years – an essay written in 1844 but never published contains all the elements of the theory – and might have waited longer had it not been for the fact that Alfred Russel Wallace, who had also read Malthus and had also studied the adaptive range of bird species in an island group, came to exactly the same conclusion in 1858. (The initial appearance of the theory was in that year, in a joint paper presented by Darwin and Wallace to the Royal Society.)

Darwin died in 1882 – seventeen years after Mendel had discovered the principles of heredity but eighteen years before they were rediscovered so that their implications could be explored. He was thus never introduced to the facts that might have permitted him to begin work revising his theory.

There remains one more aspect of the change in perspective brought about by Darwin's ideas which, though not the most obvious, is one of the most important.

During the early part of the nineteenth century the rising tide of belief in evolution rather than divine catastrophism had been tied to a growing belief in progress. In the wake of the industrial revolution the pace of change in society had accelerated, and the idea that life on Earth had some kind of direction became intellectually attractive. This notion of progress was central to the early ideas of evolution, and is clearly expressed in the basic question put by Erasmus Darwin in 1794:

Would it be too bold to imagine that in the great length of time since the earth began to exist, perhaps millions of ages before the commencement of the history of mankind ... all warm-blooded animals have arisen from one living filament, which THE GREAT FIRST CAUSE endued with animality ... possessing the faculty of continuing to improve by its own inherent activity, and of delivering down these improvements by generation to its posterity, world without end?

To Erasmus Darwin as well as to his grandson, evolution meant improvement, and this notion of progress could still embody the idea of divine purpose, progress towards a particular goal. But Charles Darwin, in discovering the mechanism of evolution, found a notion of progress that was subtly different. According to Darwin, the actual direction of evolution was determined by chance. The improvement of a species was a purely relative thing – the survivors were 'fitter' than their ancestors, but not necessarily fitter on any absolute scale, because they were not evolving *towards* anything. The notion of an evolutionary goal was not provided for by Darwin's theory. There was no sign within Darwinism of any hint of purpose.

When Lamarck had talked of evolution by adaptation he had looked at adaptation as a positive thing – a process of strategic self-improvement. Darwin's adaptation, however, was purely negative. Those who were not equipped by chance to meet the requirements of survival died, and failed to reproduce the hereditary factors which had led to their failure. This change of perspective caused a great deal of dissatisfaction among philosophers who felt that goal-less evolution was unaesthetic (and among humanitarian philosophers who thought that self-improvement ought to be worth *something* in a wider context). There was, therefore, something of a boom in neo-Lamarckism in philosophical circles, led by Henri Bergson, whose *Creative*

Evolution attempted to superimpose on the Darwinian model a new prospectus for constant improvement. Without evidence, however, his aesthetic objections were insufficient – and one suspects that if he *had* had evidence he would still have got a very poor hearing from the Darwinists, who were so sensitive to attack that any opposition was liable to be condemned as 'scientific heresy'.

The removal of any evolutionary goal from natural philosophy was important in dealing a considerable blow to anthropocentric illusions. Man could no longer be regarded as the kingpin of creation – the reason for it all. Darwin, in *The Descent of Man*, and Thomas Henry Huxley, in *Man's Place in Nature*, both took care to point out that man was not necessarily an end-product, at least in his present form. He was one stage in a continuing process, an ape with delusions of grandeur.

It was thus that Darwin paved the way for a new image of man, an image which was greatly elaborated by Sigmund Freud, who attempted to provide the first rational account of 'human nature', and to determine the structure of the human mind. The unconscious element of the mind discovered by Freud and the ape-ancestor of man confirmed by Darwin tended to come together in the popular imagination as different aspects of the same thing, as if there lurked within every man a brute beast which he was barely able to control. Robert Louis Stevenson's famous nightmare, *Dr Jekyll and Mr Hyde* (published after Darwin but before Freud), is perhaps the clearest symbolic expression of this idea.

The revelation of the animality of man, coupled with the exposure of the hopeless vanity of expecting any glorious and predetermined goal towards which evolution was directed, gave Darwin's theory an uncompromising harshness. As a picture of what went on in nature it was also harsh. Darwin himself described the mechanism of evolution as 'clumsy, wasteful, blundering, low, and horribly cruel' in a letter to James Hooker. But that, so far as he could see, was the way it was, and to deny it because it was unpleasant would be an act of great intellectual dishonesty. Many converts to Darwinism became almost proud of the harshness of the doctrine and the unflattering picture of man which it provided, considering themselves to be showing courage in accepting its implications nobly, while its opponents clung to comforting and childish beliefs. This curious brand of pride became, I think, a rather dangerous thing – particularly in its political analogues, but also because it tended to blind some scientists to the fact that there was another aspect to the survival of the fittest which Darwin had not considered at all. That, however, is something which I shall discuss in the next chapter. In the remainder of this one I want to deal with the discoveries which followed the publication of Darwin's theory, and which provide a new context for its re-examination.

Darwin referred to 'the riddle of heredity' in 1868 and admitted that biology had, as yet, no answer for it (though it was missed by Bois-Reymond and thus was conveniently unable to confound Haeckel). In actual fact, he was wrong, and the principles of heredity had already been published in an obscure scientific journal with a small local circulation in Austria. This was the result of seven years' work by an Augustinian monk, Gregor Mendel, who had performed a remarkably elegant series of experiments crossing different strains of garden peas.

Mendel's procedure was to cross two plants differing in a pair of contrasting characters and observe how the characters were reassorted in subsequent generations. This allowed him to perceive two extremely important principles.

For one thing, some hereditary characters were *dominant* over others. Thus, if a plant with red flowers (taken from a strain which had produced only red-flowered plants for many generations) crossed with a white-flowered plant (taken from a similar 'true-breeding' strain) it produced progeny which all had red flowers. All the characters studied by Mendel behaved in this way. If, however, the red flowers resulting from the cross were then crossed with each other, one quarter of their progeny turned out to have white flowers. The hereditary factor providing for white flowers had thus neither been lost nor adulterated by its association with the factor responsible for red flower colour. Mendel concluded that the hereditary factors were carried in duplicate but passed on only singly, so that the progeny of a cross received one character from each parent. Thus the true-breeding parental plants with which he had started had each had two similar 'genes'. The first generation of progeny all had two *dis*similar genes, but only one of the pair – the dominant gene – was important in determining the character of the plant. The second generation of progeny had an equal chance of getting either gene from either parent, but of the four possible combinations only one gave rise to white-flowered plants. Half the plants had one of each type of gene but were red, one quarter had two genes prescribing red flowers, and one quarter had two genes prescribing white flowers.

The other important principle which Mendel discovered was the fact that when two pairs of characters were considered together they behaved in exactly the same way *and completely independently*. Thus, if short-stemmed white-flowered plants were crossed with long-stemmed red-flowered ones, the next generation consisted of only long-stemmed red-flowered plants. The subsequent generation resulting from the crossing of these plants showed the following ratio of types: 9 long/red, 3 short/red, 3 long/white, 1 short/white. The genes prescribing white flowers and short stems had not tended to 'stick together' but had reassorted themselves quite independently.

Mendel's experiments were a model of clarity (such a model of clarity, in fact, that he has since been suspected of fudging his results to give too-exact answers) – a typical product of nineteenth-century science. It was discovered in the twentieth century, however, that things were not quite so simple. Not all genes are either dominant or recessive – sometimes an individual with two dissimilar genes actually turns out to be intermediate in terms of the affected character. Sometimes, too, genes are not reassorted randomly – some are apparently linked together and separated only rarely. In addition, many characters are not determined by the effect of *single* genes, but by the collaboration of many different gene pairs.

In spite of the many modifications which had to be made, however, Mendel's principles provided the mechanism by which individuals could pass on hereditary characteristics without their being 'blended out' by association with others, and by which a population could also retain in its gene pool a considerable potential for change, in the shape of recessive genes expressed relatively rarely only when they came together.

In 1875 Hertwig observed that fertilisation of an egg by a sperm (which had been known for almost a century to be the starting point of new life) involved the fusion of the nuclei of the sperm cell and the egg cell. It was known that the nuclei contained chromosomes, and when Mendel's findings were rediscovered the chromosomes instantly became prime suspects as carriers of the genes. Between 1890 and 1910 many close studies of the behaviour of the chromosomes during cell reproduction were carried out. It became clear that most cells contained two sets of chromosomes, but that gametes (eggs and sperms) contained only one. The single gametic chromosome set, however, is not simply one or other of the chromosome sets from a normal cell, nor even a set compiled by selecting individual chromosomes from among the double set. Each chromosome in the gametic set is, in fact, derived via a complex operation called meiosis in which the parental chromosomes join up to their counterparts and then come apart again having exchanged certain sections between themselves. Each gametic chromosome is thus a hybrid of the two parental chromosomes. The chromosomes are, in fact, chains of genes, and this is how reassortment occurs. All genes which are on the same chromosome are to some extent 'linked', and the closer they are together the more linked they are, because the probability of a 'crossing over' between the two genes at meiosis becomes very small.

The theory of genes explained how selection could operate in favouring some characters and eliminating others. But the question of where *new* genes came from remained a mystery for some time. Darwin had known that new and occasionally very strange hereditary types sometimes appeared even in true-breeding strains of domestic animals, and he called these 'sports'. He did not consider that such sports were very important in natural evolution because most of them tended to be monstrosities which died very quickly – hardly any proved viable. In 1901, however, de Vries produced a 'mutation theory' which attempted to account for the source of variability in animal populations on the basis of these sports. By 'mutations', he meant sudden gross changes in animal types – and even today, the word 'mutant' as used by writers of science fiction tends to imply something monstrous and distinct from the norm. In 1909, however, T. H. Morgan began a long study of mutations in the fruit fly *Drosophila*, and most of the innovations which he found in his populations were small ones, affecting one character at a time, usually characters determined by a single gene. Mutations in the de Vries sense tend to occur only when there is a gross aberration in the mechanism of chromosome division, and the mutation of single genes, which very rarely affects the viability of the individual more than fractionally, is a much more important source of evolutionary variation.

In the 1920s quantitative studies of mutation were mounted by Müller. He found that spontaneous mutation was a very rare occurrence, but as there were so many genes, and so many spermatozoa were formed, the number of mutations occurring per generation of *Drosophila* could be very high. Mutation could be induced by a variety of methods – certain chemicals promoted mutation rate, and it could be vastly accelerated by exposing the parent flies to ultra-violet radiation or X-rays. Müller's classic experiment demonstrating the effect of hard radiation on the germ plasm was carried out in 1927. The sun, which provides the energy source of all life on Earth, thus also provides a source of variation in the form of high-frequency radiation.

Mutation theory provided the last 'missing link' in Darwin's reasoning. If one starts at the end and reasons backwards, natural selection becomes a logical necessity. Given the nature of the hereditary mechanism and the fact that spontaneous variability does take place, then natural selection – sorting out the viable changes from the non-viable ones – is logically certain to occur. But *how powerful* would this natural selection be as an agent of change?

The answer, it seems, is: not very. The fact is that most mutations are deleterious in nature. Many of them, in fact, are potentially lethal. Most of the 'natural selection' which goes on in populations as a result of mutation and reassortment of genes consists of eliminating lethal combinations – those incapable of surviving fail to do so. These 'inevitable failures' provide a proportion of the 'excess production' of each generation – if a population did not overproduce it would inevitably decline. *Advantageous* mutations are very rare indeed, and the accumulation of advantageous mutations by a population must be an exceedingly slow process. Mutation theory thus provides very well for the 'non-survival of the unfittest', but if it is also to provide the motor for the evolution of 'superior' forms and new adaptations then the time-scale of evolution must be very large indeed.

And, of course, it is. The history of the vertebrates is not to be measured in thousands of years, but in hundreds of millions – and the history of the invertebrates goes back so far that we can no longer find the slightest trace of their beginnings. Nevertheless, the question has been raised as to whether even this time-scale is adequate to permit the mechanisms of heredity, in collaboration with random mutations, to have sieved out sufficient viable changes to account for the rich pattern of life which exists and has existed on Earth. It seems to many people that mutation must be getting a helping hand from *somewhere*, and that there must be some way that favourable mutations are 'sieved out' and harnessed which is not at the mercy of sheer blind chance. In order to investigate that question, however, it will be necessary to go much deeper into the way in which chance works, and before we do that it is necessary that we gain some insight into the arena in which evolution operates – the time-scale of the Earth and the nature of the changes which have overtaken its surface since its origin. Before discussing the process of evolution it is necessary to come to terms with the *pattern* of evolution.

6   The Pattern of Evolution

I must begin this chapter as I have begun several others by stressing the difficulties involved in obtaining and interpreting information about the history of life on Earth. The model which scientists have constructed of the evolutionary tree of life is derived from only two sources: the study of diversity of existing life-forms, and the study of the fossil record. It is important to realise that these are not *independent* sources of information.

Our interpretation of the fossil record leans very heavily on assumptions based on comparative anatomical studies of living creatures. By means of these assumptions palaeontologists find it relatively simple to infer whole structures from fragmentary finds. From a tooth, the palaeontologist is inclined to deduce a skull, from a skull a skeleton, and from half a dozen broken bones a whole sequence of species. The logic of these often sweeping generalisations is provided by the knowledge that he has acquired concerning the characteristic association of certain features and groups of features recognised today. The reverse process is also true; in trying to turn artificial classifications of extant animal groups, based on anatomical similarities, into a system of evolutionary relationships, the evidence of the fossil record is co-opted and adapted wherever possible.

In view of this mutual support we must recognise that much of the evolutionary story which scientists tend to believe is the product of their imagination. This does not necessarily imply that it is untrue, but by and large it has been put together by a process of slotting discoveries into theories rather than building theories on known facts alone. We have already noted the great importance of the creative element in twentieth-century science (and its corollary aestheticism), but there is one important difference between theoretical predictions in physical science and in palaeontology. When the meson and the neutron star became theoretically viable possibilities, atomic physicists and astronomers could go out and look for them in order to confirm the theory by observation. In palaeontology, there is no such check. We have no 'new' evidence – only evidence millions of years old. It is therefore very difficult to identify theoretical mistakes in this field.

Explorations in the microcosm and the macrocosm rely almost exclusively on inference from data which is indirect, or at least non-sensory, and yet we can still be reasonably sure of those data. The reason for this is that the data are mostly quantitative and amenable to exact mathematical description and measurement. In palaeontology, however, it is very hard to obtain good quantitative measurements. We can measure the dimensions of fossil bones without too much difficulty, but the dimensions of individual animals are not absolute specific characteristics. The measurement of one skull may give us one fact about a whole species – but it is not a fact from which we may safely deduce the characteristics of the species as a whole.

There is, in fact, only one kind of quantitative technique in the study of the fossil record which is unequivocal, and that is the method of dating rocks first used by Boltwood in 1907. Even this is somewhat limited in its applicability. The method involves investigating the extent of the decay of radioactive isotopes within the rock. From a knowledge of the half-lives of the isotopes, the age of the rocks can be calculated with some degree of accuracy. The technique was initially used with reference to uranium and thorium isotopes which decayed to lead, but these elements are not common. A method which could be applied to a wider range of rocks was not perfected until the 1960s – investigating the decay of potassium-40 into argon.

The attempt to discover the history of life on Earth on the evidence available may be compared to reproducing imaginatively the picture contained in a jigsaw puzzle when only five out of every hundred pieces are available. The method of accurate dating at least helps us to put some of the available pieces at the right height within the puzzle. We are also aided by the assumption that the completed picture looks something like a tree, and geological evidence helps us to fill in some of the background. The job would, perhaps, not be so hard if it were not for the fact that the pieces we do have are definitely not a random selection. The fossil record contains an inevitable and considerable bias.

The only animals which show up in the fossil record are those which possessed parts of the body amenable to fossilisation and also lived in places where they were liable to *be* fossilised. The best evolutionary accounts we have deal exclusively with creatures possessed of shells, scales, spines or teeth. Soft-bodied creatures have left no trace of themselves behind. Plants are known in the fossil record almost entirely through pollen grains and seed coats. Thus, though we have only five in every hundred jigsaw pieces, we find that the ones we have tend to link together. In some ways, this is convenient − it fosters the illusion that we are working well and getting the picture together. But it also means that there are vast areas of the picture which are totally blank, about which we know nothing and are not likely ever to know anything. This aspect of the problem is something most textbooks and popularisations tend to gloss over.

Even this does not exhaust the list of difficulties, for apart from all its other faults the fossil record is really only half a fossil record, or perhaps less. It begins in the middle, about six hundred million years ago. Immediately prior to that the Earth seems to have been 'scoured' − wiped clean of all traces of what went on before that date. Theories as to how this could have happened vary according to catastrophist or uniformitarian sympathies − the former favour giant tides caused by the capture of the moon, the latter a long period of glaciation. Either way, the discontinuity is there, and in trying to imagine what went on before it we have only intuition and extrapolation to help us. It is estimated that the Earth itself is somewhere between four and five thousand million years old, so that only one-seventh or one-eighth of its total history is represented in the fossil record. We have no way of knowing how long ago it was that life actually came into being on the Earth, and we can only theorise as to how.

In view of all this I think it is safe to say that there is probably no other area of science whose core of knowledge and ideas is so speculative and so dependent on assumptions whose validity is difficult to ascertain.

The age of the Earth is divided up into categories in a number of ways. The most elementary division is into eras. The time prior to the discontinuity in the fossil record is held to comprise two eras, the Archeozoic and the Proterozoic, the division between them being the time at which life originated − speculatively estimated to be about 1,600 million years ago. The time elapsed after the discontinuity is divided into three eras: the Palaeozoic (600−230 million years ago), the Mesozoic (230−65 million years ago) and the Coenozoic (65 million years ago, up to the present).

The recent eras, about which we have detailed information available, are subdivided into periods. The Palaeozoic comprises seven: the Cambrian, the Ordovician, the Silurian, the Devonian, the Mississippian, the Pennsylvanian and the Permian (though sometimes the Mississippian and the Pennsylvanian are lumped together as the Carboniferous). The Mesozoic includes three periods: The Triassic, the Jurassic and the Cretaceous. The Coenozoic has only two; the Tertiary and the Quaternary, but as these are the most recent it is again possible to make a more detailed classification into epochs. The Tertiary is thus subdivided into the Paleocene, the Eocene, the Oligocene, the Miocene and the Pliocene; while the Quaternary subdivides into the Pleistocene and Holocene.

This bewildering array of names, which have been added at various times in order to categorise prehistory by several different methods and according to several different naming conventions, is testimony of a sort to the confused history of the science. In Figures 4, 5 and 6 I have mapped out the eras, periods and epochs, and in order to give some idea of their relative time-scales by analogy I have put in parallel scales in which one day is compared to ten million years, so that the Coenozoic era fills approximately a week, time elapsed since the pre-Cambrian discontinuity fills about two months, and the entire life of the Earth is just over a year.

The surface of the Earth is made up of rocks of all these various ages, usually laid one atop another, so that outcrops of the oldest rocks are fairly scarce. Pre-Cambrian rocks are exposed in relatively few places. They are the roots of ancient, completely eroded mountain ranges. How much life there was on land in this era is not known, but it is presumed that there could have been very little, if any. The atmosphere would have been much less rich in oxygen than it is now – the oxygenation of the atmosphere has been almost exclusively the work of green plants, which produced oxygen as a by-product of photosynthesis.

The pre-Cambrian fossils which have been found include recognisable primitive plants (fungi and blue-green algae) and traces of an abundant invertebrate marine fauna. Some vermiform creatures undoubtedly existed, some shelled molluscs and some coral-like animals.

Cambrian rocks are far richer in fossil material, and there seems to have been considerable sedimentation in this period as

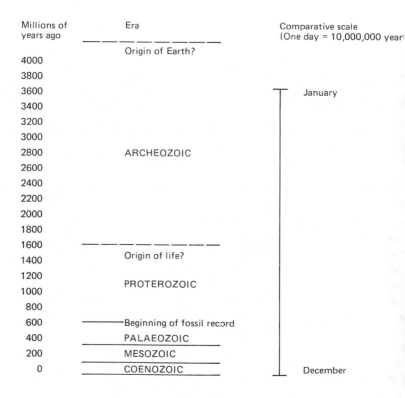

Millions of years ago	Era	Comparative scale (One day = 10,000,000 years)
	Origin of Earth?	
4000		
3800		
3600		January
3400		
3200		
3000		
2800	ARCHEOZOIC	
2600		
2400		
2200		
2000		
1800		
1600		
1400	Origin of life?	
1200		
1000	PROTEROZOIC	
800		
600	Beginning of fossil record	
400	PALAEOZOIC	
200	MESOZOIC	
0	COENOZOIC	December

Figure 4   The Eras of Prehistory

the sea covered the older rocks. The period was one of considerable geological and climatic stability, and the oceans seem to have been extremely rich in life. The most conspicuous fossils surviving from the age are the trilobites, creatures which looked rather like giant woodlice, although they were probably more closely related to present-day scorpions. Because they were so amenable to fossilisation we tend to think of them as having been the most prominent members of the fauna of the day but this is, of course, not necessarily so. They co-existed with a wide variety of sponges, many different vermiform creatures and many molluscs – some of which have cup-shaped and spiral shells not too different from modern forms. Some kinds of animals living in the sea today have analogues and/or ancestors in the Cambrian fauna, including jellyfish, sea-cucumbers and lamp-shells.

During the Ordovician period the invasion of the land began

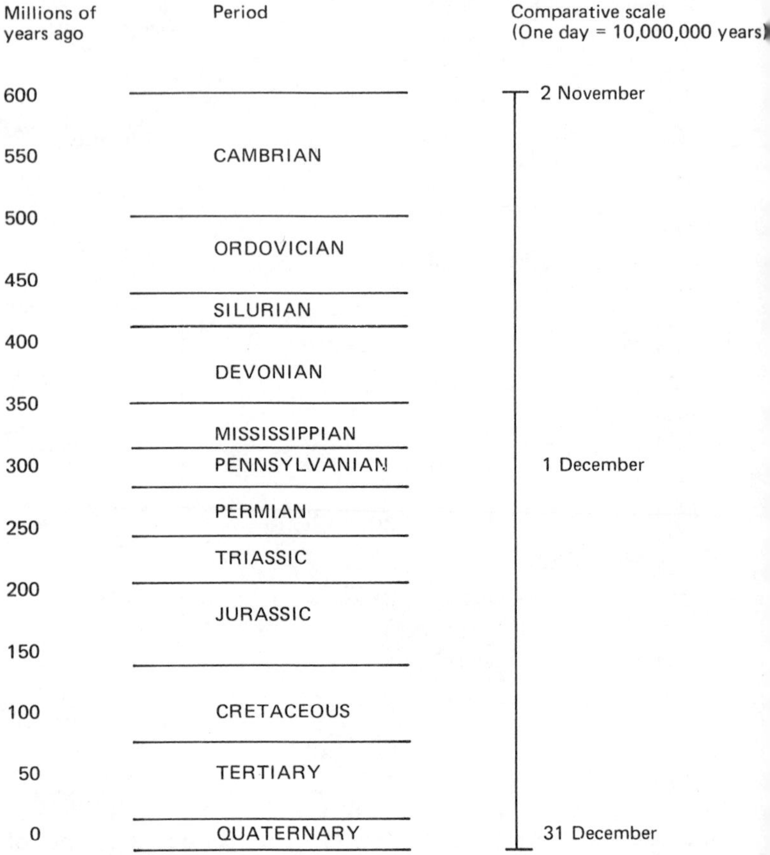

Millions of years ago	Period	Comparative scale (One day = 10,000,000 years)
600		⊤ 2 November
550	CAMBRIAN	
500		
	ORDOVICIAN	
450		
	SILURIAN	
400		
	DEVONIAN	
350		
	MISSISSIPPIAN	
300	PENNSYLVANIAN	1 December
	PERMIAN	
250		
	TRIASSIC	
200		
	JURASSIC	
150		
	CRETACEOUS	
100		
50	TERTIARY	
0	QUATERNARY	⊥ 31 December

Figure 5   Periods of Prehistory

(not necessarily for the first time) and the first traces of land plants date from this period. The trilobites were still abundant, but a new set of easily identified fossils characteristic of the age appeared – the graptolites. These were small creatures whose tubes or cups were arranged in long rows attached to floating 'buoys'. Like the trilobites, the graptolites have fallen by the evolutionary wayside and there is nothing like them extant today. Like the Cambrian, the Ordovician was a period of geological and climatic stability. Between them, these two periods lasted about 180 million years, and this long span of time allowed all the major

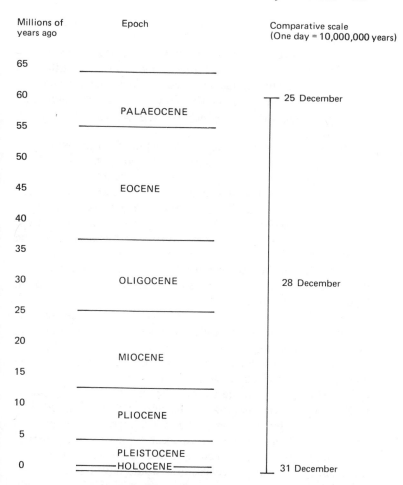

Millions of years ago	Epoch	Comparative scale (One day = 10,000,000 years)
65		
60		25 December
55	PALAEOCENE	
50		
45	EOCENE	
40		
35		
30	OLIGOCENE	28 December
25		
20		
15	MIOCENE	
10		
5	PLIOCENE	
0	PLEISTOCENE — HOLOCENE —	31 December

Figure 6   Epochs of Prehistory

phyla which are represented in the present-day life-system to become established and abundant, and to diversify slowly.

The Silurian period, which followed, brought changes. Land masses rose during this period, inland seas were formed, and the climate changed, becoming cooler over most of the surface. It is in such circumstances of change that life-forms are 'tested' by natural selection. When habitats are changing there are inevitably massive extinctions and divergent evolution becomes more rapid.

Thus, in the relatively brief time-span of the Silurian (25–30 million years) there is a considerable change in the fossil record. The fishes differentiated very rapidly while the trilobites and graptolites went into a decline. Bigger, more powerful arthropod creatures called eurypterids appeared: six-foot long scorpion-like predators with pincer-like claws. True scorpions also appeared, and are among the earliest air-breathing animals, along with millipedes.

The Devonian period, which followed the Silurian, was also a period of great unrest coupled with considerable evolutionary change. The differentiation of the primitive vertebrates begun in the Silurian continued in spectacular fashion, and fishes much more similar to modern forms appeared, as did the ancestors of the modern terrestrial vertebrates, the lobe-fins and their close relatives, the coelacanths. It was probably in this period that the first forests were established on the land, while the molluscs were still successful in the sea. The trilobites continued to decline, but their relatives, the arthropods, were establishing themselves on land quite successfully.

The Mississippian and the Pennsylvanian make up the Carboniferous period – so called because this was the time that the great primeval forests and swamps laid down the coal measures we exploit today. During this period changes in climate and the elevation/submersion of the land apparently became cyclic. Thus, although the overall climate was relatively stable, local circumstances may have been very turbulent indeed. The same *range* of habitats was probably preserved throughout this period, but there would have been constant geographical redistribution of them. Such changes must have been important in the evolutionary pattern because they would have favoured those animals which were *behaviourally* adaptable: able to cope with change or to migrate. In this period, therefore, we can detect little change in the population of the sea, save that trends already established continued. On land, however, things were different, and a rich land fauna developed from the arthropods and the amphibians. By the later stages the first true reptiles had evolved and the familiar types of modern insect – flies, cockroaches, etc. – were apparently common.

The evolution of the reptiles from the amphibians was a particularly important step in the evolutionary sequence, because the reptiles were the first vertebrates to gain independence from the sea. They laid eggs with shells, which were resistant to desiccation and could be laid on land. The evolution of this 'cleidoic egg' was a direct response to the cyclic changes taking place in the Carboniferous and which continued through the Permian period which brought the Palaeozoic era to a close. Without such a regime of seesaw changes the selective pressure to become independent of water would not have been anywhere near so strong.

During the Permian the cyclic changes came to a stop. The continental land-masses rose never to sink again and other land-masses sank never to re-appear. The changes in climate culminated in a series of glaciations. The end of the cycle brought about a wave of extinctions in the land-fauna as there was a dramatic 'weeding out' of those species which could not cope with the new regime. It was after this complete overhaul that the biosphere took on the aspect which it retains to the present (save for the diversification of the 'higher' vertebrates from the 'stem-reptiles' which emerged in the period to dominate the vertebrate fauna). In the sea, the trilobites finally bowed out, following the graptolites and the eurypterids into oblivion.

A certain stability returned to the world during the first period of the Mesozoic – the Triassic. After the extinctions of the Permian (which were probably coupled with a severe reduction in numbers within many populations which did survive) the relative stability allowed those groups which had stood the test to consolidate their position and to begin again the process of slow specialisation which had been interrupted by the regime of unrest favouring more versatile types. Inevitably, the greatest capacity for specialisation was contained within those groups which had been the most versatile, and Triassic saw the beginning of the adaptive radiation of the reptiles, which led to the evolution of the dinosaurs and the mammals.

The giant dinosaurs which we know so well appeared in the Jurassic and persisted through the Cretaceous. Dinosaurs and prehistory have become virtually synonymous in the popular imagination. The late Mesozoic is often described as the age 'when dinosaurs ruled the Earth', and one of the most impressive carnivores of the period has been officially credited with the name *Tyrannosaurus rex*, 'king of the thunder-lizards'. The dinosaurs are by far and away the most spectacular group in the fossil record, and they are now completely gone. Because we tend to associate size and power with success the question of 'why the dinosaurs died out' seems to us to be something of a mystery. (I know of one attempt to rewrite the story of prehistory whose first chapter is provocatively entitled 'Who shot the dinosaurs?') But we must remember the inherent bias in the fossil record. *Prominence* does not necessarily imply *dominance*. Size and power are not synonymous with evolutionary success – rather the reverse, in fact, as very many species surviving today are the economy-size models whose giant cousins could not stand the pace. The only close relative of the dinosaurs which has stood the test of time is the tuatara – a small burrowing lizard living on islands near New Zealand.

It was not just the dinosaurs which were subjected to the great wave of extinction which happened at the end of the Cretaceous, but many other groups as well. Again, the end of an era brought profound changes in the biosphere. Curiously, however, the Cretaceous does not seem to have been a time of great geological upheaval or climatic change. How, then, were new conditions generated to test the specialist species which had emerged during the Mesozoic and find them wanting? I have already pointed out the important difference between sequential changes in climate and changes in the inherent variability of the *weather* within a climate, and it seems likely that what took place in the late Cretaceous was a steady increase in the instability of climatic conditions. The cause of this change was probably the drifting apart of the continents.

The land-masses we know today were initially elevated at the end of the Carboniferous and the beginning of the Permian. At that time, however, there were only two vast continents, one in each hemisphere. The Northern one, generally termed Laurasia, consisted of North America, Europe, Greenland and Northern Asia, while the southern one – Gondwanaland – was an amalgam of Africa, South America, Australia, Antarctica and India. They were separated by a narrow strait, and the rest of the globe was covered by a single great ocean – the Tethys Sea.

The continents began to drift apart in the early Mesozoic. The cause of the break-up is unknown – catastrophists have suggested meteor-strikes or the capture of the moon, but uniformitarians favour the theory that the initial agglomeration was unstable anyhow, and had to come apart in order to redistribute the Earth's mass more equably. The drifting of the continents was very slow, but the cutting up of the land-masses and their encirclement with ocean must have brought about quite crucial changes in local climatic conditions. When there were just two vast land-masses the relative dearth of coastline must have meant that conditions on the continents followed a reasonably simple pattern, with a fairly smooth geographical distribution of habitats. But when the break-up was complete (when Africa and South America split apart in the late Cretaceous) this was no longer true. The drifting of the continents continued, of course – India and North Asia did not collide until the mid-Coenozoic, and Africa did not rotate to seal off the Mediterranean until later still – but the essential separation would have made all the difference to the orderly distribution of local climates. It was this which probably caused the 'testing' of the Mesozoic fauna. That testing had precisely the same result as the similar selective regime in the Permian: the specialist species tended to die off while the most versatile survived. In the first instance the versatile reptiles had out-competed the amphibians, now the most versatile product of the reptile-adaptive radiation – the mammals – survived while the specialist reptiles went to the wall.

The drifting of the continents, incidentally, was only confirmed in the early 1960s although Wegener's studies of prehistoric climates had led him to the conclusion in 1911. Wegener was regarded as a crank for a long time until new evidence was obtained from the alignment of magnetic ores in rocks laid down in different periods of prehistory. These magnetic ores had been laid down in a direction determined by the magnetic poles; and showed how the various areas had been orientated towards the poles at the time of their origin, and from their distribution the drifting of the continents could be tracked through prehistory with some accuracy.

The Coenozoic era again saw the restoration of slow sequential change in local conditions, and the pattern of the Mesozoic was repeated as the mammals underwent considerable adaptive radiation throughout the Tertiary period. The 'Age of Reptiles' had lasted 200 million years, and had given rise to twenty groups known in the hierarchical classification as 'orders'. The 'Age of Mammals' – which is, of course, still with us – has lasted only one third as long and has given rise to thirty distinct orders. This testifies to the greater versatility of the mammal species, which came into their own at the beginning of the Coenozoic – greater versatility permits greater opportunity for re-specialisation when the regime of change alters to permit adaptive radiation rather than selective testing.

In the Tertiary the ancestors of the mammals of today first

differentiated and arrived at the basic 'blueprints for survival' – the various forms which were to prove 'profitable' in the evolutionary sense. To a large extent they simply repeated the same specialist patterns the dinosaurs had found: the rhinoceros is an analogue of the ceratopsian dinosaur, the elephant of the sauropod, the tiger of the carnosaur, the dolphin of the ichthyosaur etc., etc. The same ways of life became practical again, and the mammals evolved to re-fill the 'ecological niches' which the dinosaurs had failed to hold during the period of stormy weather. The fossil record of the Coenozoic, looked at a hundred million years from now, might suggest that the tiger and the elephant 'ruled the Earth', and the fact that they did not, in the long run, prove to be an evolutionary success might appear to be a problem: why did the giant mammals die out? But we know perfectly well why the giant mammals failed (and are failing) – they could not compete with the more versatile types, the non-specialist species, when conditions changed so that competition became acute.

It is not known exactly when the ape-like ancestor of man became sufficiently distinct from the similar creatures which were the ancestors of the modern great apes to be identified as the base of our particular branch on the evolutionary tree. The first appearance in the fossil record of a species which seems to have been more like man than the modern apes is in the Miocene epoch, two-thirds of the way through the Tertiary period – about 20 million years ago. This species, *Ramapithecus*, is known to us only through a few teeth, and we know very little about it. He may, however, have been a common ancestor of man and the great apes.

We know little about the evolution of hominid primates in the next 10 million years – half the time-span which separates us from *Ramapithecus*. Some fossils found in coal mines in Italy apparently bear on the problem, but the descriptions of them we have are mostly second-hand. A few teeth and jawbones dating back 10 million years, and named *Oreopithecus* may represent the human branch, but it is also possible that they are common ancestors of men and apes.

Most of what we know about the prehistory of man is based on skulls and bones found in Africa. In 1924 an important skull, which seemed to have several features in common with human skulls but not represented in ape skulls, was found in South Africa and named *Australopithecus*. Further investigation turned up other similar skulls, and abundant jaws and teeth. In 1959 skulls of this type were shown to exist at the same geological level as primitive stone tools and flakes. The shapes and sizes of the skulls show a fairly considerable variation. Potassium-argon dating has established the age of some of the skulls at about 1,750,000 years, though most are more recent. This places *Australopithecus* in the Pleistocene epoch of the Quaternary period.

Just as the final periods of the other eras saw the end of a long regime of slow change in a flurry of rapid cyclic change – an increase in climatic variability rather than gross climatic change – so the Quaternary period brought a new change-regime to the Coenozoic in the form of the Pleistocene Ice Ages. During the Pleistocene glaciers extended themselves from the poles deep into the temperate regions four times, interrupted by interglacial periods when local climates were probably very similar to today's. The present day is very probably an interglacial period – there is no reason to expect that the glaciers have retired for good. The last Ice Age (in the Carboniferous and Permian) lasted far longer than the Pleistocene one has – so far. In identifying the last few thousand years as the beginning of a new epoch, the Holocene, we are therefore being rather optimistic.

As in the Permian and the Cretaceous, the new regime of change began to sort out the products of the mammalian adaptive radiation. A new wave of extinctions began (and is still with us, although we cannot perceive it because of our different time-scale), and once again it was the specialist species – the ones which lacked the behavioural adaptability to cope with variable circumstances – which suffered. The Pleistocene extinctions have often been associated with the rise of man and his exploits as a hunter, but this is unlikely. In fact, the Pleistocene extinctions and the emergence of man probably have a common cause – they are different aspects of the same selective process. In the Permian the stem-reptiles had been the evolutionary success-story. In the Cretaceous it was the early mammals. In the Pleistocene it was a certain type of mammal – man.

In every case it was the big and the strong who failed to compete while the small, adaptable types survived. In every shake-up the meek inherited the Earth – no matter what they went on to become in the periods of adaptive radiation which followed.

There is some doubt as to where, exactly, the Australopithecine ape-man evolved to become the genus *Homo*. Some proponents of an 'African genesis' credit the African Australopithecines with being the direct ancestors of man because Africa remained untouched by the glaciations, and this seems quite likely. But the idea that man survived by 'hiding out' from the regime of change in Africa does not really leave room for an explanation of the rapid evolution which went on in the human branch of the evolutionary tree during this period. Various 'advanced' fossil men have turned up in Africa, but they have also turned up in many other parts of the world, in areas which were covered, at intervals, by the ice. The type of skull which has been called 'Neanderthal man' and skulls much more reminiscent of modern man have been associated with both the first and second interglacial periods in Europe. It is not entirely clear whether modern man and Neanderthal man were really separate species or whether the skulls represent two extremes of a continuum. What is clear, however, is that the evolution of *Homo sapiens* from the Australopithecines did not take place entirely in Africa: it seems more probable that the African hominids migrated north as the glaciers retreated during each interglacial period, and were thus kept under environmental pressure which forced them to remain adaptable. It is possible that Neanderthal man and some of the most recent Australopithecines were variants which tried to specialise – one by avoiding the ice and one by co-existing with it – and that both failed to compete with the more versatile ancestors of *Homo sapiens*, but this is pure conjecture. At any rate, in the third interglacial period (30,000 years ago) modern man seems to have established himself as the sole representative of the genus. The change in man's physical form since that time has been almost negligible.

Having completed this brief sketch of the pattern of evolution (and not forgetting that it is a very tentative sketch), it is now time to ask whether we can find, within the pattern, some indication of the kind of process which is taking place.

We have seen that the tree of life does not branch at random. What we find, in fact, is that clusters of branches emerge during periods of relative stability, as the trunk of the tree splits many ways. Ultimately, however, the most divergent of the radiant branches tend to reach dead ends and a new radiation takes place from a new main stem. Each new main stem tends to have certain things in common with its predecessors; it is the branch which has not grown away from the tree into the divergent paths of extreme specialisation. The mainstream of evolution, as far as we can track it back, consists of the creatures which have kept their evolutionary options open and kept their potential variability in reserve. The most impressive creatures in the fossil record – the trilobites, the eurypterids, the giant dinosaurs, the sabre-toothed tigers etc. – have not been part of that mainstream, but merely unsuccessful experiments.

But why? Darwin's concept of natural selection involved the survival of the fittest. According to Darwin, it was the large and the strong and the well-armed which survived while the small, the weak and the defenceless failed to compete. The fact that this is not the case does *not* imply that the theory of natural selection is wrong, but that Darwin's idea of the way natural selection operates in nature was mistaken. For various reasons already discussed the implications of the theory were never worked out logically, and Darwin put the emphasis in entirely the wrong place. Survival of the fittest does *not* mean the survival of the biggest, fastest, most effective killers.

In a sense, it would have been far better if the word 'fittest' had never been used at all, for fitness implies to us bodily strength and speed – the fitness of the athlete who wins races and contests of strength. But in the context of natural selection fitness means something very different. Natural selection is 'survival of the most efficient' not 'survival of the most effective'.

The difference between being 'efficient' and being 'effective' is subtle, but extremely important. The most efficient predator is not the most effective – the *effective* wolf is an expert sheep-killer who can slaughter his prey at will, but the *efficient* wolf is the wolf who does not slaughter his prey at will, but makes sure that there are always enough sheep left to thrive. In the evolutionary sense the successful predator is the predator which conserves its prey for future generations; the better the sheep do in the game of natural selection the better the wolves can do.

Natural selection operates within the struggle for existence, but the struggle for existence is not like a knock-out competition where everybody fights a long sequence of battles and only the one team that wins them all survives. The struggle for existence is a matter of maintaining a delicate balance in which a whole group of teams must maintain their integral relationships.

Wolves eat sheep and sheep eat grass, and the three co-exist in mutual support. What is good for the grass is good for the sheep, and what is good for the sheep is good for the wolves. But the whole thing must stay balanced. If the wolves kill too many sheep, then the rate of conversion of grass-into-sheep-into-wolf declines because there are not enough sheep. On the other hand, if the wolves do not kill enough sheep, then the rate of conversion of solar-energy-into-grass-into sheep declines because the grass is over-cropped.

Now, in order to maintain the stability of the relationship some form of 'negative feedback' is necessary. (In a way, it is a similar situation to that encountered in a previous chapter with respect to a main-sequence star where gravitational force and the expansion of the solar nucleus counterbalance one another, and any increase in one means an increase in the other, causing balance to be restored.) Such negative feedback generally operates in the short term. When the wolf population increases the sheep population decreases until a threshold is crossed and the wolves face starvation, at which point they begin to decline and the sheep begin to increase in consequence. These 'predator–prey cycles' are known to occur in certain instances in nature, and are generally helped by the fact that the balance involves not three elements but four. The carnivores are generally limited by parasites, so that when the carnivore population begins to increase the rate of increase of the parasites speeds up and helps cut back the population before the prey population is threatened with extinction. Also assisting this 'balance of nature' is the fact that there is usually more than one species at each rank in the hierarchy, so that as one declines to the point at which it is threatened with extinction the pressure upon it is relaxed because the predator switches to an alternative prey. (In the parasite/host cycle this does not tend to happen, but here there are genuine feedback cycles operating because of the thresholds at which parasite transmission becomes so easy that an epidemic breaks out, or so statistically unlikely that a healthy population can build itself up again.)

A large number of factors help to sustain the balance of nature in the short term – that is to say, from one generation to the next – but it is always a slightly uneasy balance, for if it ever tips far enough to knock one element out of the structure altogether the whole thing may come tumbling down – and the only ones who survive will be those who can adapt to a new way of life.

The value of adaptability is therefore quite clear. But if versatility is so valuable, why do specialists evolve at all? If, in the long run, it is always uneconomical to be big, strong and effective, why do creatures with these characteristics evolve constantly?

This is a new question, but it is, I think, the right question to ask. The enigma of the dinosaurs is not a question of why they died out, but of why they ever evolved in the first place. They died out because in the struggle for existence they had evolved to become effective and not to become efficient. They were, in fact, 'unfit' in the broad sense of the word, and yet they had been favoured in selection over a long period of time, within a special kind of change-regime.

The giant dinosaurs – and all the other spectacular failures of the fossil record – were victims of what I shall call a 'growth-trap'. It is the sort of trap that can develop only in times of stable or sequentially changing conditions, when a sufficient number of generations of a species live in the same habitat, with the same way of life.

The basic situation, in its simplest form, is the wolf/sheep situation already described. The wolf population and the sheep population are balanced, with feedback ironing out any considerable changes which take place in their numbers in the short term. It is within this situation of balance, over a considerable period of time, that natural selection begins to have an effect.

It is in the nature of the situation that some sheep must die in every generation to feed the wolves. These sheep will tend to be the weaker, slower sheep, and so the sheep, on average, will tend to get quicker, stronger, and may grow horns or armour in order to help them withstand the attacks of the wolves. The inherent variability of the population is there already – it only needs the selective pressure of the wolves and a long, long time to bring it into play. However, we must also look at the situation from the viewpoint of the wolves. Some of the wolves in each generation will also die, especially when the sheep population is on the decline. The wolves which live will tend to be the quicker, stronger wolves, least daunted by the horns or armour which natural selection is giving the sheep. On average, therefore, as the sheep become better wolf-avoiders, the wolves will become better sheep-catchers. Here we see Darwin's notion of constant 'improvement' of the stock of each species. In reality, there is no change in the 'balance of power'. Predator and prey evolve in parallel. But this change is *directional*.

Let us imagine a hypothetical predator and its prey. Each has a certain limited capacity for variation, and it so happens that natural selection in each case favours the same kind of change. The prey species is being slowly modified so that it is becoming more agile – better equipped to run away from the predator. As it does so, it exerts selective pressure on the predator to become more agile itself, to become better at running after its prey. The evolution of the predator will thus increase the selective pressure on the prey to become faster still, and this pressure will cause evolution in the prey which will exert selective pressure on the predator to become even faster.

Now this is not negative feedback at all, but positive feedback. The faster the prey becomes the faster the predator becomes and so *ad infinitum*. The problem is, of course, that such directional change cannot proceed *ad infinitum*. There comes a point where the adaptability of the body to the production of further agility is simply no longer possible mechanically.

The only way out of the positive feedback cycle is for the one species or the other – usually the predator – to find an alternative way of going about things. But 'finding an alternative' is a matter of pure chance; if there is no alternative available within the variability of the species then none will be found, and predator and prey will be trapped in the cycle, with neither able to break out. This commitment may in itself be a 'bad thing' if conditions are changing rapidly. But in a situation which is stable over a long period of time, changing slowly and smoothly, then the time becomes available for the feedback to carry on – not *ad infinitum*, but *ad absurdum*.

These specialist characters always occur in pairs. As a prey species develops its armour, a predator develops its methods of armour penetration. As a prey species grows big and strong, its predator grows big and strong. As a prey species develops its defensive weaponry, its predator develops its counter-defensive weaponry. And, when the time-span permits, these changes go on to their ultimate forms – animals so bulky and so magnificently-toothed that they can hardly carry their bodies and their heads. When the limits of mechanical possibility are reached there is no further scope for change, and the feedback process ends – but by then there is no hope whatsoever for the paired species. They cannot hope to compete with the more adaptable forms, and it takes only a spell of bad weather, or even a statistical accident upsetting their balance of numbers, to wipe them out.

Evolution is not reversible. Selective pressure causes

directional change if it is maintained for any length of time, and a species caught in a singular sequence of directional change is caught in the jaws of a slowly closing trap, for the sustained and steady pressure must persist until no further change in that direction is possible, at which point the species must decline and die out under the same pressure. The survivors – the ones who do not get caught in such traps – are the species which do not become subject to directional pressure, the species which, when times get hard, can adapt themselves behaviourally rather than bodily. There is an old proverb to the effect that he who fights and runs away lives to fight another day. In the evolutionary context this should read: He who fights and runs away lives *only* to fight another day. Species which are successful in the struggle for existence conducted under the rules of natural selection are species which keep both fighting and running away to a minimum. In the evolutionary story, an ounce of adaptability has always been worth a ton of brute strength.

We see, therefore, that success in the struggle for existence is, in the long term, not quite what is implied by the phrase 'survival of the fittest'. But we are still looking at only one side of the coin, for there is far more to survival than the struggle for existence.

For a character to be favoured by natural selection the gene or genes prescribing that character must increase within the gene pool of a species from one generation to the next. Darwin saw this increase being achieved by negative means – that is to say, individuals carrying alternative, less favourable genes died and were unable to contribute to the gene pool of the next generation. But such an increase can also be brought about by positive means – by the individuals carrying the genes being able to make a greater contribution to the next generation in terms of viable offspring.

This does not simply mean producing *more* offspring, for producing more offspring generally means producing weaker offspring. It means, essentially, *optimising* reproductive effort to produce the maximum number of offspring which is compatible with imparting to them the equipment necessary to their own survival.

There are thus *two* kinds of natural selection – the subtractive selection envisaged by Darwin, in which the natural death quota of each generation tends to remove 'poorer' genes, and the additive selection of increased reproductive effect. (We may, for convenience, label the first 'latent environmental subtractive selection' or LESS, and the second 'maximization of reproductive effect' or MORE.)

It must be stressed that with respect to these two classes of selective pressure we are dealing with two different classes of genes. Genes controlled by LESS are genes which prescribe physical characters – not just strength, speed and shape but also the ability to use particular types of food, the capacity for self-repair in case of disease or damage, the capacity to withstand temperature changes etc. These genes are concerned with the efficiency of the bodily machine in sustaining itself. The kind of genes governed by MORE are genes which provide for the organism to make the best use of its potential as a reproducer of its own kind. Only a few of these genes are concerned with mechanical factors (i.e. the simple production of eggs) – most of them are concerned with behavioural characteristics providing for better parental care of eggs and offspring.

We have seen that subtractive selection is an evolutionary force of somewhat ambiguous effect. In certain circumstances it leads species not towards 'immortality' but into traps and dead ends. In terms of the tree of life subtractive selection is a force which produces an abundance of dead branches. But this is not true of MORE. Here there is no delicate balance of nature, no positive feedback. Evolution in this direction – towards better parental care – is always constructive. It is here, if at all, that the real capacity for improvement enters into the evolutionary process. It is this class of factors which provides for evolution as a whole to be directional, and in a sense, even provides evolutionary goals.

In the early part of this chapter we saw a pattern in the evolutionary story by which the adaptive radiation of one class of animals came to a dead end in a wave of extinctions, and then a new class began a new adaptive radiation. The amphibians were succeeded by the reptiles, who were succeeded by the mammals, whose adaptive radiation is also failing while the least-specialised member of the class thrives. But this is not just a cyclic process, an evolutionary seesaw which goes nowhere, for as each new adaptive radiation begins there *is* 'something new under the sun'. Each new class is always possessed of something its predecessor lacked: *a new means of securing the future of its offspring.*

The clumsy and wasteful kind of natural selection which Darwin saw is the kind which produces giant dinosaurs and hopeless growth traps. But behind the pattern of explosion and extinction there is a slower evolution going on; one which is not 'clumsy, wasteful, blundering, low and horribly cruel'. This is the process of evolution, which is an equally inevitable logical consequence of the mechanisms which permit natural selection to occur, that lies unheeded while the ideological battle goes on.

We can see now a new significance in the evolutionary story. The advantage which the reptiles had over the amphibians which were before them was the cleidoic egg – essentially, the *eggshell*. All the physical differences between the reptiles and the amphibians proved to be unimportant – they duplicated most of the physical types during their own adaptive radiation just as the mammals duplicated all the reptile types in theirs, and they proved no better for such increases in mechanical efficiency as there were. The vital difference was that the reptile egg was better provided for than the amphibian egg. The eggshell is the first step in the maximisation of reproductive effect among the vertebrates.

Reptiles generally do little more for their offspring than give them a slightly better start in life, but their close relatives the birds do much better. The birds are a kind of side-branch on the evolutionary tree which first became distinct in *Archaeopteryx* and *Hesperornis* of the Cretaceous and survived in parallel with the mammals. Birds almost invariably look after their young for some time after birth, working hard to supply them with food. It is in the birds that we see the first instances of behaviour-patterns designed for the protection of the young from predators – the well-known strategy employed by some ground-nesting birds where the parent 'pretends' injury in order to lure a predator in pursuit of her and away from the nest is a good example. But the birds were not the success story of the Coenozoic, because the mammals did it all so much better.

The reptiles protected their eggs by putting shells round them. The mammals protected theirs by not laying them, but keeping them wrapped up inside the body and giving birth to live young. The birds fed their young by catching food, but the mammals actually made feeding the young part of the ordinary function of the organism. The birds initiated protective behaviour – the beginnings of 'maternal instinct' – but the mammals took it much further. Maternal behaviour in the mammals has become very complex and very various.

Here we come to the crucial point in the story: the evolution of man, and in particular, the evolution of the mind. What factors were important in the evolution of man from the animals? What, in the evolutionary context, are the brain and the mind for?

Those committed to the 'hard' Darwinian viewpoint have argued that man emerged triumphant from the pattern of evolution because he was the fittest – the most efficient hunter, the most efficient competitor, champion of the world. They have argued that intelligence evolved to provide man with advantages in the struggle for existence; they would argue that the crucial points in human evolution were the discovery of the club and the spear. (In a sequence of the film *2001*, which purports to show the dawning of intelligence in a community of primeval apes, the crucial incident is when one ape takes up a bone and starts using it to smash other bones and – later – to smash other apes.)

There is no doubt that the evolution of intelligence permitted the invention of weapons, and provided strategies for their use. But if the factors important in exerting selective pressure on the pre-human apes had been the necessities of attack and defence then the apes would have evolved claws and fangs, not brains. The factors which were important in exerting the selective pressure which led to the evolution of intelligence were the factors involved with *better parental care of offspring*. The evolution of the brain is connected to the evolution of more complex and more adaptable behaviour patterns concerned with the rearing of children.

The record of life on Earth testifies quite clearly to one thing: that the race is *not* to the swift, nor the battle to the strong. The species that we glorify – the lion (king of the beasts), the eagle (king of the air), the king cobra and *Tyrannosaurus rex* – may seem to us to be powerful and lordly, but in the evolutionary context they are inconsequential aberrations. The rat and the sparrow have far more hope of a long evolutionary life-span than the lion and the eagle. (But then, royalty is passing out of fashion in the human world, too.)

The image of man which Darwinism promoted in the popular imagination – the image of Mr Hyde in Stevenson's story – is a distorted image. From the chapters which follow, I hope we may distil from an examination of some twentieth-century discoveries in the biological sciences a much better understanding of 'human nature'.

7

The Chicken and the Egg

In this chapter I want to approach the subject of evolution from a new angle. In the last chapter I tried to provide some insight into the process of evolution by describing the pattern we find in the history of life so far as we can ascertain it. In this chapter I want to try to gain some insight *via* the discoveries which have been made concerning the nature of life and its organisation.

In the *Origin of Species* Darwin did not go so far as to argue that life had initially evolved from non-living material – he dealt only with how living forms changed. Nevertheless, the implication was present in his theory, and it was only a few years after the debate began that Tyndall spoke at the British Association of the possibility that life had emerged within the 'primeval slime' by purely natural processes.

Not long before Darwin life had been considered to have a special chemistry of its own. Although living tissue was known to be made up of the same elements as ordinary matter, the actual compounds concerned seemed to be unique to organic matter and could not be synthesised from common substances. When, in 1828, Wöhler first synthesised urea in the laboratory a minor controversy was provoked. A link between the common chemistry of physical science and the sacred substance of life was a threat to the Cartesian partition, which was already suffering under the attack of the evolutionists. Wöhler's achievement was undermined by the allegation that some of the material used in his synthesis was, in fact, of living origin, and that he had not really succeeded in making an organic substance from inorganic material. It was not until thirty years later, in the late 1850s, that Berthelot managed to synthesise some compounds of low molecular weight (formic acid, methane and acetylene) and confirmed the link between the two chemistries once and for all. The real breakthrough in organic chemistry, however, was theoretical rather than practical – the discovery of structural formulae pioneered by Kekulé's imaginative model of the hexagonal benzene ring.

Just as Dalton's theory had given chemists formularistic command over inorganic chemistry, Kekulé's inspiration paved the way for the formularisation of organic chemistry, and it was this imaginative control – the ability to quantify – which really offered hope for a rational chemistry of life. This discovery was made in the 1860s, just a few years after Darwin's theory was published.

The first practical application of the new theory of organic compounds was in the dye industry. In 1856 Perkin, tinkering with organic reactions in the hope of synthesising quinine had discovered a purple dye by accident – the first synthetic aniline dye. Perkin became rich and tinkering with organic materials became fashionable. Chemists in Germany used the new methods of formularisation to rationalise and control the reactions which they were exploring, and new compounds were soon being identified and isolated by the score. (In 1875 Perkin, tired of making money, went back to dabbling and synthesised coumarin, thus founding the perfume industry as well as the dye industry.)

The making of artificial organic compounds went on in parallel with the attempt to analyse the essential compounds of life, but this latter research went much more slowly, for the simple reason that the essential structures of living tissue tend to be very complex molecules indeed. The structural formula of glucose, the simplest of the carbohydrate sugars, was finally worked out in 1886. The amino acids – building blocks of the proteins – were separately identified and their formulae determined in the same period, although the nineteenth and last (of the most important and commonly occurring ones) was not discovered until 1935.

The discovery that most of the important structures of life were made up of long chains of simpler chemical units was a biochemical revelation akin to the other great conceptual simplifications of the nineteenth century. It was found that compounds like cellulose and starch were gigantic molecules made up of simple sugars strung together like beads on a chain. The proteins, which were the actual 'fabric' of life, were long chains of amino acids folded into complex three-dimensional shapes. Building the compounds of life, therefore, was not simply a matter of synthesising organic compounds from inorganic ones, but a matter of putting together the units in the right sequence to make the appropriate chains.

The magnitude of this task was not fully appreciated until well into the twentieth century. Fischer, in 1907, managed to string a few glycine units together to make a 'peptide chain', but until the Second World War no one managed to make a chain with more than nineteen units. An average-size protein used in living tissue contains about six hundred amino acids, strung together in an absolutely specific order, and folded just right. It was not until 1953 that du Vigneaud actually succeeded in building a functional protein in the laboratory – the hormone oxytocin, which consists of only eight amino acids.

In seeking to provide a model for the origin of life from inorganic materials, therefore, biologists were faced with several very awkward problems. Firstly, if organic compounds were so difficult to synthesise from common molecules, how did the chemicals of life – especially the amino acids – originate? Secondly, once amino acids *did* exist, how was the business of building them into the complex and multifarious proteins, each one specifically designed for a particular role within a complex system of functions, begun, controlled and perpetuated?

In the 1920s, the Russian biochemist Oparin published a classic work which undertook to explain how reactions occurring in the primeval atmosphere of the Earth could give rise to an abundance of organic compounds. He assumed that in the ancient atmosphere there was no oxygen or nitrogen (both having been produced by the activity of living plants) and that it was, in fact a 'reducing atmosphere' akin to the atmospheres of some of the other planets in the solar system, containing hydrocarbons like methane and acetylene, and also ammonia. These compounds reacted with water – or, to be more accurate, with superheated steam, as the Earth's surface temperature must then have been far higher than it is today – to give a complex range of organic acids, alcohols and aldehydes. The formation of the amino acids is only one step removed from this kind of conversion.

To what extent the Oparin hypothesis may be true is not certain. Notions about the compositions of the primordial atmosphere remain very largely conjectural. In a long series of experiments dealing with different versions of such a primitive atmosphere, however, Miller and Urey succeeded in generating complex organic compounds – including amino acids – by passing electrical discharges through mixtures of gases including ammonia and methane.

It may be possible, then, to imagine that as the Earth cooled and its water vapour condensed into a warm ocean the sea was really a hot dilute 'soup' of organic compounds. What kind of things might happen within such a soup which could permit the beginning of the second stage of organisation?

It is presumed that this 'soup' was chemically active, with new compounds being formed continually. Many – especially the larger molecules – would not be water-soluble but would exist in a colloidal state. Even in solution complex molecules would not tend to be evenly distributed but would, via a process known as coacervation, become associated in limited regions.

In 1958 Sidney Fox formed compounds consisting of short, amino-acid chains from inorganic materials and dissolved them in hot water. As the solution cooled the 'protenoids' formed coacervate globules which he called microspheres. The microspheres were not simply passive bundles of molecules – active chemical processes were going on within them. Such microspheres, formed in the primeval soup, may have been the ancestors of living cells.

Microspheres can 'feed' and grow by absorbing chemicals selectively from the dilute situation in which they are suspended. They may divide in two – 'reproducing' by fragmentation. (The process of the condensation of microspheres out of chemical soup is analogous to the condensation of stars out of interstellar gas, save that the progressive organisation of a star is cohesive, while that of microspheres is replicative. The star gets ever greater while the microspheres become ever more numerous, but both processes involve the spontaneous concentration of material from the environment.)

The cycle of progressive growth-and-reproduction is the basic property of life, although it is not entirely confined to living systems. The probability of such self-replicating systems appearing by pure chance, even under the most favourable environmental circumstances, is undoubtedly very small – but the probability is finite. The essential thing is that once it *has* happened, the probability of its being 'unhappened' by pure chance becomes very low, because of the property of self-replication. Any chemical environment conducive to the formation of microspheres, given enough time, is likely to evolve systems of self-replicating bodies, which may begin to 'compete' for the available chemicals. Natural selection may thus have begun before the first living cell was formed.

The process of microsphere formation in chemical soup is

essentially one of concentrating chemical functions rather than the chemicals themselves, and the process by which microspheres might evolve towards living organisms is one of organising chemical functions. When we look at the processes involved in the replication of the simplest living organisms extant today we may be struck by the extreme complexity of the nucleic acid molecules and the incredible delicacy of the enzyme-regulation system by which it is controlled. It seems virtually inconceivable that such molecules could have evolved to perform their functions purely by chance. But we must remember that it was through evolution of the functions that the molecules came to be made, and not vice versa. The nucleic acid replication system, though now universal, may have originated at the end of an extremely long evolutionary chain of such replicatory processes as the most efficient (the 'fittest') applicable to the environmental circumstances of the time.

There are two vital steps by which microspheres might organise themselves into cells. The first is the formation of a boundary – a 'skin'. This step is probably quite an easy one – we are familiar with numerous processes in which the interface between two chemical systems forms a solid boundary (as, for instance, when hot milk cools). A skin formed at the coacervate/solvent interface is almost certainly the most primitive form of cell membrane.

The only well-defined structure in the most primitive cells we know is the outer cell membrane. (Most bacteria have a cell wall as well, but this is not always the case.) Within it, however, a further step in organisation is required as special functions within the cell have to be localised – a kind of 'division of chemical labour'. This may be accomplished by a furthering of the process of coacervation within the outer skin, so that microspheres-within-microspheres are formed. Some of these, too, may acquire the property of self-replication. All of this adds a new level of complexity to the proto-cell. In so far as physical structure is concerned, bacteria and some other forms of primitive living organisms – notably the blue-green algae – are really no more complicated than this, and we may imagine that the first life-forms on Earth had very much the same kind of physical appearance. Such evolution as there has been between the first cells of all and modern bacterial cells is chemical; the refinement of replicatory processes.

The business of cell-replication is conducted in the nucleus of the cell. There, molecular blueprints are carried as extremely long molecules of nucleic acid – usually DNA (deoxyribonucleic acid) but sometimes, in very primitive cells, RNA (ribonucleic acid). Normally, the DNA is organised in a double helix, which might best be envisaged as a twisted zip-fastener. Only one half of the zip's 'teeth' is blueprint – the other half is a complementary 'template'. The teeth are chemical bases, of which there are four: adenine, guanine, cytosine and thymine. Each of these can only link up to its chemical partner – adenine to thymine and cytosine to guanine, so that if the chemical sequence of bases on one half of the zip is AGCATGCTTAG then the sequence on the other half must be TCGTACGAATC.

When the two elements of the double helix are 'unzipped', each half collects new chemical material to build a new opposite half, so that the blueprint side builds a new template and the template side builds a new blueprint. In this way, the helix replicates itself.

But the nuclear material can do more than replicate itself, for the sequence of bases on the blueprint half of the molecule is a 'genetic code' for building proteins. Each group of three bases corresponds to an amino acid, and a sequence of several hundred bases corresponds to a particular amino-acid chain: a protein or part of a protein.

Thus, as well as building a new template half for itself, the blueprint half of the double helix may build a different counterpart molecule of 'messenger-RNA'. This messenger molecule may then migrate out of the nucleus to another area of the cell, where it builds a new blueprint molecule out of another species of RNA called 'transfer-RNA'. The new blueprint molecule then gathers amino acids from its immediate chemical environment and strings them together to make the proteins whose specifications are written out in code in the DNA.

The elaboration of the entire process by which this complex process of chemical organisation is conducted and regulated, and the subsequent 'cracking' of the genetic code, which took place during the 1950s and 1960s, is one of the most impressive achievements of twentieth-century analytical science.

At the moment, it is presumed that, in the vast chromosomal strings of the nucleus, each gene codes for a single protein – an enzyme. Each enzyme controls, by acting as an intermediary, some chemical reaction which goes on in the cell. It is by enzymes that the replication of DNA, the making of messenger-RNA, the making of transfer-RNA, and the gathering of amino acids into proteins are 'supervised' and regulated. Thus the process by which DNA makes enzymes which allow the making of DNA and more enzymes is a circular one – a closed cycle. Once such a cycle was initially sealed, in the remote past, to become self-supporting, it could proceed by a kind of 'chemical perpetual motion', energised by the fixation of solar energy through photosynthesis, evolving systematically from relative chemical simplicity to the bewildering complexity of the contemporary system.

The gross functioning of a primitive cell may be compared to a factory. Molecules taken in from outside have to undergo long, elaborate sequences of chemical processing, being initially dismantled or re-shaped into usable components and then assembled again into functional molecules – some of which are used structurally, others as enzymes to control the process. The way in which factories have evolved, first to the mass-production of components and then to the assembly line, and finally to the factory-complex, is a reasonable analogue of the way cells must have evolved initially. A modern car factory, where the product passes along a conveyor belt, with groups of men at strategic points adding pieces, is organised according to the same principles as the cell manufactory where proteins are built.

All this, however, is only half the story of the evolution of life on Earth. This account allows for the evolution of bacteria-like entities from non-living material, and it is predominantly a matter of chemical evolution. But once chemical evolution reaches a particular stage of efficient organisation, then a different kind of evolution can begin – the evolution of variant physical forms, of organisms.

The need for compartmentalisation of the elementary cell arises initially from the need to accommodate a larger number of chemical operations. For a time, primitive life-forms in the primeval soup would be involved in the processing of amino acids which existed freely in the environment, but after a time the availability of such material would decline as it became incorporated into the life-system. As the free amino acids declined, speciation – the divergent evolution of specialised cells within the life-system – would begin. There would be selective pressure on the primitive cells to develop characteristics facilitating one of two new lifestyles; with amino acids no longer free in the environment, the cells would either have to make their own out of smaller molecules (i.e. become plants) or steal them from other cells (i.e. become animals).

Even in the most primitive life-forms this differentiation of way of life is possible − the blue-green algae are simple molecule-builders, while most bacteria are simple molecule-stealers. However, the complication of the kind of operation which goes on within cells puts a high premium on structural modification for efficiency (just as the layout of operations in factories has to be changed occasionally in the light of 'time-and-motion' studies). Thus the compartmented *eukaryot* cell evolved from the uncompartmented *prokaryot* bacterial cell.

The compartmentalisation of the eukaryot cell was accomplished by a great elaboration of the cell membrane, which folded inwards until it filled the cell rather like a loosely crumpled ball of material, with a single fold wrapped round the outside and another inner fold wrapped around the nucleus of the cell. Within the folds pieces of material became pinched off to form small packets of cytoplasm called *organelles*, which specialised in various chemical operations essential to the cell. Chloroplasts are plant-cell organelles which handle photosynthesis − the building of organic molecules from atmospheric carbon dioxide and water, using solar energy. Mitochondria are organelles occurring in all cells which handle the 'respiratory function', processing organic molecules − fats and carbohydrates − and making energy-rich molecules which provide 'fuel' for the nucleic acid replication and protein-building functions. Other organelles include lysosomes − bags of enzymes which, if liberated, dissolve and destroy the cell.

There are many organisms which consist of single eukaryot cells specialised in some particular fashion: the protozoans, including the shape-changing *Amoeba*, the multiciliate *Para-mecium*, and *Euglena*; and the simple algae such as *Chlamydomonas*. All these cells are built to the same pattern of organisation, but are modified for various means of locomotion and various ways of life. Perhaps the most interesting are some species of *Euglena* which can grow chloroplasts and live as plants when light energy is available, or abandon them and live as animals when it is not, thus preserving the most elementary of their options.

Most of these primitive eukaryots replicate by straightforward binary fission (or 'multiply by division' as a slightly ironic description has it). This simple process is the basis of the fundamental property of life − reproduction. There is, however, an alternative process – or, more accurately, a complication of it – whose origins probably go back to the chemistry of microspheres. Microspheres tend to grow and divide, but occasionally they also tend to meet and merge, becoming overlarge in the process so that they quickly redivide. In the process, however, they exchange some of their material, and therefore the cells which result from the subsequent divisions will be 'hybrids' of the two original 'parent' cells which merged. This property of merging, reassortment of materials, and division must have been crucial to early evolutionary processes in recombining chemical characteristics so that the best combinations could be sorted out by natural selection. When prokaryot cells first formed they retained, to some extent, this property of combining and reassorting their constitutional material, and quite complicated procedures evolved by which the single chromosome of the prokaryot cell could link up with another chromosome and exchange genetic material. It is a process which bacteria indulge in relatively rarely, and many species extant today do not seem to go in for it at all, but it was a process that was retained through natural selection as an important method of shuffling characteristics; successful innovations tended to have come about through this process because it is thus that potentially beneficial mutations are able to find their appropriate context. The successful innovations, of course, would tend to retain the capacity for further recombination.

The process is known as sexual reproduction, and even at this primitive level it involves a separation of individuals within a species according to the role played in the process. Bacteria do not 'merge' in the sense that microspheres do – one bacterium 'injects' its chromosome into the other, so that there is a genetic donor (a 'male') and a genetic recipient (a 'female'). There is, however, no sexual differentiation in bacteria – an individual might adopt either role.

In eukaryot cells, which are structurally more complex, the property of sexual reproduction also has to become more complex, and the mechanical difficulties necessitate quite elaborate processes. But because this facility is one of the things which allows evolution to take place, evolution inevitably selects the most efficient means by which sexual reassortment can take place. In a sense, the whole history of evolution from the first eukaryot cells is the history of the evolution of better and better methods of achieving sexual reassortment of genetic material.

Classifications of the life-system as it now appears to us often make a basic intellectual division between 'one-celled organisms' and 'many-celled organisms', lumping the prokaryots and the primitive eukaryots together in the first category and all more complex forms in the latter. This division, however, is a false one. 'Many-celled organisms' are merely a phase which one-celled organisms go through.

All life-forms on Earth are basically one-cell creatures. Some eukaryot cells, however, have evolved inordinately complex systems of sexual reproduction which involve the growth of many-celled structures. (Some have also evolved inordinately complex systems of asexual reproduction which involve the growth of many-celled structures, but this is probably a by-product of the processes of natural selection favouring more efficient means of sexual reassortment.)

We ourselves are many-celled entities, and with our customary egocentricity we regard our many-celled aspect as the *product* of evolution and the egg-and-sperm system as the *means* by which reproduction occurs. From an objective viewpoint, however, we can see that the human being is merely the means to an end: a reproductive mechanism. It is the single cell, the egg, which is the basic entity, and what we think of as the organism is merely a special structure built by the egg according to specifications contained within itself in order to reproduce itself.

There is an old joke to the effect that a chicken is only an egg's way of making another egg. This is, in fact, the correct perspective from which to look at the evolution of life. It is the single cell — the egg — which is the basis of all life on Earth, and multi-celled entities are simply variant forms of reproductive apparatus. The answer to the conundrum about whether the chicken or the egg came first is easily answered — the egg came first and foremost, and the chicken is only one in a long series of experiments which eggs have carried out in order to discover the most efficient means of making more eggs.

With this perspective it is much easier to understand the pattern of evolution which we investigated in the last chapter. The vital importance of the cleidoic egg becomes much easier to see when we realise that the struggle for existence is really fought out between eggs, and the 'fittest' are those eggs which increase their own numbers at the expense of other kinds. It was not so much that the reptiles thrived because they invented a better egg, but that the reptile eggs thrived because they had invented a better kind of egg-maker.

The mammals, seen from the same viewpoint, thrived because the mammalian eggs abandoned their shell-makers and invested in incubators instead. The human egg emerged triumphant among the mammal eggs because it invented the most efficient incubator – a fully automatic, self-servicing, decision-making incubator, which was even – to a certain extent – self-regulating too.

It is even possible to answer, in a fashion which is only slightly flippant, the question of *why* we are here on Earth. We are here because a pregnant woman is the ultimate logical development of the eggshell.

The development by eggs of multi-cellular reproductive apparatus permits further division of labour and structuralisation of organisation. The protozoan cell is an all-purpose cell which carries out all the essential functions of life: intake of materials from the environment, processing of those materials, growth, excretion of chemical waste generated in the processing, and replication. The metazoan egg, however, can delegate most of these functions to its reproductive apparatus, which, being multicellular, may develop special multi-cellular structures – organs – to cope with each function individually. If a cell may be analogised to a factory, therefore, the multi-cellular corpus may be likened to the whole of a nation's co-ordinated industrial effort. Virtually all the cells in a multi-cellular entity are, however, short-term investments, born to serve their purpose and then die. Only the reproductive cells – the eggs and their counterparts, the sperm – live on beyond the death of the whole apparatus, and it is their structure and capabilities which are the real subject of evolutionary change.

The range and diversity of the cells which exist on Earth is tremendous, even when the specialised cells with built-in obsolescence are set aside. The smallest life-units (that is to say, the smallest known bacteria) may contain 800 million molecules. An amoeba, a fairly simple all-purpose cell, may have 2×10^{17} molecules – nearly a billion times as many. Size, however, is not necessarily everything. Though the human egg cell contains more molecules than an amoeba, the human sperm cell – which contains a complete set of the genetic information required for building a human being – only contains about 7×10^{11} molecules (one millionth the number contained in an amoeba.) On the other hand, the human egg cell is not very large as metazoan egg cells go, because it does not have to store a great deal of the sustenance required to feed a developing embryo – most of that is supplied by the maternal cells. A chicken's egg is therefore 35 million times as big as a human egg.

The diversity of entities in the life-system goes beyond cells, however. Bacteria are the simplest organisms which exist but they are not the simplest organic entities. Viruses may be smaller and much simpler. Viruses do not have the power to replicate themselves or to grow, but they do have the ability to subvert the natural processes which go on in living cells and force the cells to make viruses rather than more cells. Viruses are thus a rather strange kind of predator; instead of stealing organic molecules from cells to use as raw material in their own factories they actually take over the factories of cells and use them to produce their own products.

Viruses actually consist of bits of chromosomal material wrapped up in protein coats. The chromosome does not contain enough genetic material to code for the complex business of self-replication, but when joined on to another chromosome in a host cell it can use the host's replication-genes in association with its own blueprint-genes. Whether viruses originated as fragments of cellular chromosomes, or whether they were initially bacterial cells which lost their ability to self-replicate in discovering a new mode of parasitic existence, remains an open question.

The smallest viruses – the smallest entities in the life-system – are no more than complex molecules, containing no more than a few thousand *atoms*. The largest, on the other hand, are of a size with the smallest bacteria. The viruses may represent evolution in the 'reverse direction' – towards structural and chemical simplicity rather than complexity. Their existence is, however, dependent upon the initial evolution of living cells.

Viruses are a kind of 'inbuilt-disaster' within the life-system: a by-product of life which actually threatens the existence of life. Like the growth-traps to which multi-cellular organisms are prey, this is one of the hazards of natural selection – a 'fault' in the progressive system. Another such fault is cancer, which arises as a fault in the egg's organism-building system. It can be induced chemically or by viruses, but what actually takes place is that the genetic message controlling local cell-division and differentiation somehow becomes garbled: cell-growth runs wild, completely out of control. Finding treatments for cancer is so very difficult because it is not a specific 'disease' with a singular cause, but a general kind of organic malfunction which may have many causes in the 'thousand natural shocks that flesh is heir to'.

The fact that one cell containing about the same amount of matter as a pinhead can contain within it sufficient molecules complicated enough to specify the building and functioning of an apparatus as complex as a human being is really quite astonishing. It is difficult to conceive that such a monumentally accomplished egg could evolve from a bacterial cell by chance alone. But chance, of course, does not operate alone and unconstrained. It has been said that a few monkeys given typewriters and allowed to bash away at random would, in the fullness of time, eventually produce the complete works of Shakespeare. But the bacterial cell, mutating away at random and reshuffling its attributes occasionally via sexual reproduction, has a helping hand that the monkeys do not have, for only the products which are successful stay in the cycle. Every mistake which happens is immediately thrown away. If there were some such selective agent working with the monkeys, which meant that only the real words they produced could be repeated, and that the real sentences they formed were likely to be formed again while every nonsensical one was not, they would produce the complete works of Shakespeare so much faster – perhaps, on the time-scale of the universe, quite quickly.

The fact is that the more complex the reproductive apparatus whose blueprint is contained within the egg, the more severe the regime of natural selection becomes. Blueprints which are liable to go wrong, or produce inefficient products, are eliminated from the life-system almost at once. It is at this level that subtractive selection is most important – *not* at the level where the pieces of apparatus compete among themselves. In evolution, the 'fittest' are not the strongest body-machines but the cleverest eggs, which contain the best recipes for efficient egg-makers. In a sense, Darwin was right, and evolution does consist very largely of the weakest going to the wall – but the weakest which fail are the eggs which never develop and the embryos which abort of their own accord. Any creature which breaks out of its eggshell or crawls out of its womb has already survived the most rigorous subtractive selection.

Having made these few brief points about the nature of life in general, I now want to move on to a special case: the strategies employed by one particular egg – the human egg. I am tempted to introduce it as 'the most successful egg of all', but I shall refrain from such blatant anthropocentricity on the grounds that time has not fully tested the experiment, and it may yet be that the humble bacterial cells which have never even gone in for the most elementary structuralisation will outlive the whole category of life to which we belong. Nevertheless, the reproductive apparatus employed by the human egg *seems* to be markedly more efficient than all other products – it seems, in fact, to have removed itself on to a whole new plane of efficiency and even to have exempted itself from the dictates of natural selection. I think it is a fair question, therefore, to ask what is so special about the reproductive apparatus of the human egg, and why there are facets to its character which do not appear in the physically similar products of other, closely related eggs.

The human being has been described as a 'naked ape', but this is a description which prejudges the question of what kind of being man might be. There is a world of implication in choosing to label man a 'naked ape' rather than a 'big-headed ape' or an 'upstanding ape' or an 'adaptable-handed ape'. In a sense, it is prejudicial to describe man as any kind of an ape, in that this implies that we know exactly where apes fit into the general pattern of life, and only have to modify that knowledge slightly in order to figure out what kind of being man is. This is just not so. I think that in order to gain any real insight into the nature of the human being we must adopt – initially, at least – a perspective which contains far fewer 'hidden assumptions': a viewpoint which is a little more remote.

Let us, therefore, begin by asking elementary questions about how man compares to the whole spectrum of the vertebrates, of which he is a relatively recent product.

The size spectrum of the vertebrates spans about eleven orders of magnitude – from tiny fish weighing a fraction of a gram to whales weighing more than a hundred thousand kilograms. On land a couple of orders of magnitude are missing at each end, so that whether we consider the whole range or only the range of land vertebrates the middle range is the same – between 10^3 and 10^4 g. An adult human may weigh about a hundred kilograms (10^5 g) and may be said to be above average in size, though he remains two orders of magnitude smaller than giants like the elephant and *Tyrannosaurus rex*.

Man's size, then, is not particularly remarkable. His shape is slightly more so. He is bipedal – a characteristic which he shares with such oddities as the ostrich, the kangaroo and *Tyrannosaurus rex*, though he walks more upright than any of them. He has long arms, like the great apes, but he does not use them either for walking on or for swinging from. His mechanical versatility is thus much greater than other creatures adopting a similar stance. He is not particularly remarkable in using tools – thrushes use anvils and sea otters use levers, but he is remarkable in the wide use which he can make of them, in the adaptability of his hand to gripping and manipulating. Apart from this handiness his body is not particularly able – there are many creatures which are either stronger, more agile or better armoured than he, and some are better in more than one respect. (Few, however, have his all-round competence in these matters.)

Perhaps the most remarkable quality of man's physical make-up is its durability. This may, of course, go hand in hand with his all-purpose design, for specialisation in strength, speed or armour undoubtedly puts mechanical strain on a body. The life-expectancy of modern man is over sixty years, and even in poor circumstances such as prevail in the worst environments it is over twenty. In the Stone Age it was probably nowhere near as great – perhaps the average life-expectancy of a new-born child could be measured in months. But it is not the average life-expectancy which we are measuring in considering the durability of the human apparatus, so much as the time that a man *can* live if he survives to the age at which he becomes self-sufficient. This is usually reckoned as 'three-score years and ten', although only in modern circumstances is that total likely to be attained by the majority.

Among the other mammals only whales and elephants may reach such a prodigious age, although some reptiles and the odd bird or two have been credited with great antiquity. (Most stories of the great age achieved by certain animals are probably apocryphal, but even if we accept them they must be very exceptional individuals indeed.) But actual length of life is not really a good measure of durability, for different creatures live their lives at different rates. The metabolic rate of an elephant is considerably slower than that of a man, and the metabolism of a shrew very much faster. A good guide to the 'tempo of life' is the heartbeat, and so if we calculate life-spans in terms of the number of heartbeats a body generally lasts, we have numbers which are at least roughly comparable. If we do this we find that the shrew and the elephant both live 'as long' as one another – about one billion heartbeats. Virtually all other mammals vary from this figure by less than 10 per cent. The variance among cold-blooded creatures and birds is rather greater, but for the most part their 'ration of life' is far less. No other vertebrate begins to approach the heartbeat record set by man, whose three-score years and ten contain nearly three billion heartbeats. We may say, therefore, that the human body, despite all its failings, is about three times as durable as any other animal body – given, that is, that we look after it as well as we are capable.

Why should this be so? More important, perhaps, why should this be an evolutionary advantage? For the egg, it must be remembered, builds bodies only to make more eggs, and makes them with built-in obsolescence. The answer to this question is, I think, the key to the true specialness of the human being.

The human being is born at a very early stage in its development, and spends a long time growing to maturity. This characteristic, known as neoteny, can be very useful to organisms under certain circumstances. Certain amphibians – notably the axolotl – practise neoteny as a means of coping with the possible range of their environments (i.e. as a means of preserving versatility). The axolotl is basically a large tadpole, but unlike most tadpoles it has the option of remaining one, growing reproductive apparatus and reproducing as a tadpole without ever bothering to metamorphose into the 'adult' form – a salamander. When water is abundant it breeds as an axolotl, but whenever its habitat dries up it can change into a salamander capable of withstanding a considerable amount of drought and desiccation. This element of choice is most important in securing the future of the species in conditions which could not normally allow amphibians to survive.

Neoteny in humans is not so simple. Protracted youth in humans is not a facultative thing but an inevitable one. From egg to adult the human egg takes sixteen or seventeen years to develop, only nine months of which is spent within the womb. During the greater part of its development phase, therefore, the developing human is amenable to the influence of the environment – within the limited plasticity prescribed by its genetic complement it is allowed to adapt to the specific circumstances in which it finds itself. This is true of humans to a far greater extent than it is true of other animals.

Does this protracted youth help to explain the durability of the adult body? If the human body were apt to wear out after the usual billion heartbeats, it would do so after about twenty-five years, thus giving the human animal twice as much youth as adult life. This is not, in itself, particularly unusual. There are many creatures which live only a very short time as adults after a long youth – the mayfly is perhaps the cardinal example, spending a whole year as a larva and only a few hours as an adult. But a mayfly adult lives only to breed. It has no other function (some adult insects are not even equipped with mouthparts or digestive tracts, only with reproductive organs and the locomotory apparatus necessary to locate a mate). In the case of mammalian adults in general, this is not true. The mammal adult has *two* functions to perform in carrying out the business of the mammal egg – to make more eggs *and to ensure that those eggs develop.*

The protracted adulthood of the human animal now becomes understandable. The adult body has to be durable in order to keep itself going long enough to provide the appropriate measure of care and attention for the human young. Whereas the adults of most non-mammalian species become redundant soon after breeding, the adults of mammalian species only become redundant after they have exhausted their potential for parental care. The young of the human species require parental care for a much longer period than other mammalian young, and thus the life-span of the human being has to be considerably longer than the life-span of other mammals.

This gives us a superficial answer to the question. As with all answers, however, it merely reveals deeper questions. Neotenic mammals require long adult life-spans. But what is the advantage in being neotenic? Why is it such an advantage to the human egg to prolong the development of its reproductive apparatus so much? We have already touched upon the answer to this question, which is quite simply that neoteny permits options to be kept open.

We have already seen, in the previous chapter, how important the preservation of versatility is within the pattern of evolution. At every crisis point in evolutionary history, the specialist species have died in droves while the life-forms which possessed behavioural adaptability survived. Even in the simple form in which it is manifest in the axolotl, neoteny promotes behavioural adaptation – it preserves a choice of ways of life absolutely necessary to an organism living in a capricious environment.

Human neoteny preserves the plasticity of the developing human being through long years, allowing the maximum opportunity for the growing human to adapt behaviourally to its environment. Instead of its behaviour being ruled by instinctive behaviour-patterns controlled by its genes, it is able to *learn* behaviour-patterns appropriate to specific circumstances.

This ability is present to some degree in virtually all mammals and some birds. The victory of the mammals over the reptiles was the triumph of parental care over the eggshell (and that of the birds – a more limited victory – was the triumph of parental care plus eggshell over eggshell alone). With parental care comes, as an inevitable corollary, the capacity for parental guidance; the possibility that what parents learn about their environment can be passed on to their offspring.

The exploitation of this capacity through the accumulation of chance mutations has been very slow, but we can see its effects throughout the range of the birds and mammals. The existence of this possibility made the development of elaborate communications systems in mammal and bird species the subject of selective pressure; the evolution of such systems became inevitable. And evolve they did, though we are only just beginning to find out about them. (The complexities of bird song and the language of the dolphins have only been investigated in recent years, and scientists have confessed great surprise at their findings. Birds and mammals communicate much more effectively that we would – until recently – have imagined.)

The enhancement of behavioural adaptability and the efficiency of communication required considerable evolution to take place with respect to one particular organ in the mammal/bird reproductive apparatus: the brain. The brain controls behaviour – it is the computer which runs the body-machine, processing incoming sensory data and converting them into information about the environment, and then prescribing the appropriate motor responses.

The capacity for the development of parental guidance in individual species was therefore governed by the development of the brain. In order for the new ability of the mammals and birds to be fully exploited the brain had to become bigger (so that it could carry and process much more information) and more elaborate (so that its processing could become much more sophisticated). The extent to which the brain could develop within the mechanical limits allowed by the structural forms of the various species determined the extent to which parental guidance could be exploited as an evolutionary strategy.

The birds were extremely limited in the amount of brain-expansion which could take place within their bodily framework. There was simply no way that a bird could carry a big head – no possible evolutionary sequence of form could allow the kind of body-machine possessed by the birds to turn into the kind of body-machine necessary to accommodate a big brain. With the mammals it was different. It must be remembered, however, that after the Cretaceous extinctions the mammals were under selective pressures of all kinds. As their adaptive radiation began, the physical forms of most families were growth-trapped into evolutionary sequences which set strict limitations on brain size.

Most mammal species were committed by their particular selection regimes to four-footedness, and that commitment effectively ruled out the development of very large and complex brains; there was no way they could be made mechanically sound. Only two kinds of mammal retained the essential capacity for making body-machines mechanically capable of supporting large brains: those which retreated to the sea, where weight is so much less important; and those which remained on land and developed an upright stature, with the skull supported on a strong spinal column. So it was, in fact, inevitable that the most intelligent species on Earth should arise among the whales (the so-called 'killer whale' and dolphins) and among the primates (the chimpanzee and the human).

I think that we can now understand the specialness of man, in all its aspects. Without the evolutionary sequence which led to a large brain, the evolutionary sequence leading to extreme neoteny would not have been forced by selective pressure, and vice versa. These characteristics evolved in association, by a kind of feedback process not too dissimilar to that which operates in growth-traps. In a sense, it is a growth-trap, for the human brain cannot simply get bigger forever. If natural selection *continued* to operate within the feedback context then we would, like the big dinosaurs, be evolving slowly towards mechanical impracticality. (Many science fiction writers used to imagine just that – they foresaw a future in which the man of the year million was all head and no body, and became extinct through mechanical inefficiency.)

But this cannot happen, for the simple reason that natural selection cannot and does not continue to operate in such a situation, as it does in a growth-trap. The other half of the feedback system, which favours the evolution of communicative methods and parental guidance of the young, is not a mechanical factor at all, and its end-product is not a machine pushed to the very limits of practical design, but something altogether different: a mind. This particular process of evolutionary feedback has not taken the species into a dead end. It has, instead, taken the species out of the context of natural selection altogether. It has given that species the ability to implement its own regime of *un*natural selection, and to take control of evolution away from pure chance and into its own hands.

Human beings exist today – and have existed for many thousands of years – in a regime of evolutionary change in which chance and natural selection play virtually no part. The biological evolution of the human species carried that species over a critical threshold (perhaps, even without egocentricity, we may describe it as the most critical of all the evolutionary thresholds) beyond which a new kind of evolution – social evolution – took over as the dominant force of change within the species.

When Darwin published his theory of evolution, he borrowed phrases and assumptions from students of social evolution, and students of social evolution borrowed from him. In the late nineteenth century it was widely believed that the principles governing social evolution were similar to, if not the same as, the principles governing biological evolution. This was perhaps the worst mistake the rationalist philosophers made, and it is certainly the one which has had the most tragic consequences in terms of its practical applications.

In this chapter I have given an account of the biological processes which were important in the evolution of man from the 'primordial slime'. In the next I shall endeavour to show how fundamentally different are the processes which now control evolution within the species.

8   Human Evolution

I have characterised the human animal as the animal with the greatest capacity for behavioural adaptation; the animal whose behavioural patterns are no longer subject to the dictates of chance but to the dictates of *choice.*

What the human body-machine has that the machines produced by all other eggs have not is the ability to select from a wide range of alternative reactions to particular stimuli; it has the most flexible genetic programming. Evolutionary commitment almost invariably leads to extinction – only lack of commitment has long-term survival value on the evolutionary time-scale. The human being is the least committed of all creatures.

The privileges of choice are easy to see, for they are the privileges that we enjoy. The ability of man to master even more complex and powerful tools – both mechanical tools and intellectual tools – and materially to alter his environment to suit his needs and desires, in fact everything which has converted human evolution into human history, is the result of the ability to choose and to discover new choices. But what is the mechanism of choice? How, in evolving from ape to man, did man acquire choice?

The reactions of an animal are controlled almost exclusively by instinct. A primitive mammal has all its decisions programmed into it: behavioural patterns are linked to a series of perceptive triggers. When a particular pattern of sensory information is collated in the brain, a sequence of switches is closed and a pattern of motor responses is initiated. The first question we must ask in seeking the origin of choice (and, incidentally, attempting to answer the seventh of Bois-Reymond's world-enigmas – the one which even Haeckel declined to take on) is: how do alternative courses of action become incorporated into an instinct-ruled system of behaviour-determination?

There is a very ancient thought experiment which involves

placing an ass exactly equidistant from two exactly identical bundles of food. The idea is that because the two stimuli are identical, but operate in different directions, there would be no way the ass could choose between going one way or the other, and would die by starvation thanks to its inability to make a decision. This is, in fact, an extremely unlikely set-up, but there is another story (perhaps anecdotal) regarding a method of catching monkeys by exploiting their inability to make 'rational' decisions. This method involves putting a banana inside a trap, whose entrance is wide enough to permit the monkey to reach in, but not wide enough to allow the hand clutching the banana to be withdrawn. Once the monkey has the banana, it is alleged, it is obliged by its behavioural programming to hang on to it, and thus cannot escape, even when the trapper returns to take possession of his captive.

In order to 'escape' from these predicaments two distinct kinds of choice are required. The ass has to decide between two identical but opposing stimuli. The monkey's decision is harder – the urge to seize food being naturally greater than the urge to avoid capture he has to upset his own genetic programming. These examples illustrate the two kinds of situation which require that some degree of choice be incorporated into the determination of behaviour.

In the first instance, the agent of choice is simply probability. The smallest difference in the stimuli affecting the ass will tip the decision balance one way or another; if one bundle of food looks slightly bigger, or the distance slightly less, then the decision is made. The essential thing to note is that even if no such difference exists in the framing of the experiment (which, being a thought experiment, is not limited by practical considerations) the ass may still make a decision because of differences in its own perception of the situation. The experiment may be perfectly designed but the ass is not – it sees the world with slightly prejudiced eyes. (Nowhere in nature is there a perfectly symmetrical entity.) In a fifty-fifty situation, therefore, decisions can be taken on the basis of chance – a 'left-prejudiced' ass will go left, and vice versa. There is no deadlock.

The monkey's problem cannot be so simply solved. Here we are dealing with two different stimuli, of which the 'wrong one' is determining the monkey's behaviour. This is the more interesting situation, for the problem here – from the evolutionary standpoint – is how does the monkey rearrange its own built-in priorities so that it can release the banana and run away? What is needed is some kind of *motivational control*; it must be able to adapt to its particular situation by temporarily re-ordering the hierarchy of its motives.

It is this second kind of choice which is important in human evolution – *not* the kind of probability-determined choice-making which influences the ass. (Haeckel, in calling Bois-Reymond's last enigma an illusion, was assuming that all choices are made on the probability-determined basis – he was, however, being rather too narrow-minded).

How, then, can such motivational control evolve? How did the proto-human ape evolve a motivational system which allowed him to adapt his instinctive motive patterns to specific situations?

The answer is that in the human species a new series of motives appeared which could at least modify, and in some circumstances overrule, the instinctive motivational system.

The motive forces of instinct are quite simple – they are the priorities of evolution. The function of the egg's reproductive machine is to reproduce, and therefore it is designed to sustain itself (by supplying itself with food and water) and to produce more viable eggs (by breeding and rearing young). Primitive mammals, therefore, have four 'basic' motivational forces: to eat, to drink, to copulate and to protect the young. The human species, under selective pressure to improve its versatility, had to evolve not only *more* motivational forces, but an *extra system* of motives which could co-ordinate and control the basic one.

These new motives did not evolve at random. The motives which evolved had to be the ones which gave individual humans a genetic advantage over others within the species. The way to secure such a genetic advantage was, of course, to be more successful in rearing offspring. (An alternative means would have been to murder the offspring of other humans, but that would be a much less efficient way of increasing one's own genetic contribution to the next generation – as an evolutionary policy it is quite self-defeating, and with infant mortality as high as it must have been in proto-human groups, there must have been much more capacity for improvement through preservation than through destruction).

The new motivational system which evolved, therefore, comprised motives useful in securing the future of the young. Any motive force which added to the efficiency of child-rearing would have been favoured by selective pressure. Logically, one is forced to conclude that the most basic component of the new motivational system – the one which would provide the most considerable evolutionary advantage – would have to be the motive force which we experience as the emotion of love (not sexual love, but protective love). In the conditions which presumably existed in the 'society' of the pre-human apes, there can have been no motive force so powerful in enhancing the chances of any infant's survival to adulthood – through a long neotenic period – as maternal love.

It is important that we understand the essential difference between the motivational forces of the secondary system and those incorporated into the primary one. The first set of motivational forces are instinctive in nature, programming specific behaviour-patterns in response to specific stimulus patterns. The secondary system evolved as a means of strategically upsetting the strict hierarchy of the instinctive behaviour, making behaviour flexible by making the instincts less effective and the whole behaviour-determining system inefficient (or even 'irrational' where the strict evolutionary priorities are assumed as the standard of rationality).

The secondary motives thus tend to be anything but specific: they cannot and do not manifest themselves through simple stimulus/reaction patterns, but in a generalised sense, as feelings. Their function is not to facilitate fast, automatic action, but to slow down action responses, to create doubt, to let the experiences of learning manufacture alternatives and allow flexible choice-priorities to be invoked.

This is a difficult point to grasp, and I think we must appreciate the distinction between *motives* and *emotions*. Emotions are the way in which motivational forces are consciously experienced and expressed. Motives themselves are, however, independent of consciousness. The motivational forces of the primary system can operate quite without reference to consciousness (and probably do in virtually all non-human mammals) – they are merely connections in the brain. The secondary motivational system, however, requires the conscious mind as an arena of operation. The capacity to transcend unconscious instinctive programming originated via the process of developing and operating a conscious mind, in which motivational forces could be expressed and compared as feelings. The conscious mind thus evolved as the overseer of the unconscious, to oppose and control it.

It has sometimes been argued that the conscious mind arose as the seat of reason, but this is not so. We can only build a competent model of the evolution of the human brain if we assume that the conscious mind first arose as the seat of the emotions – the conscious aspects of motivational forces. The critical threshold crossed by the human species when the dictatorship of chance gave way to the dictatorship of choice was not – as many writers of prehistoric romances have assumed – the birth of reason but the birth of emotion. Choices are not necessarily dictated by reason (as, surely, we have all observed).

The emotions made it possible – ultimately – for human motives to become partially free of instinctive programming (made it possible, in fact, for motives to become 'mixed'). In the matter of its basic design, however, the secondary motivational system remained under the control of natural selection. The decline of natural selection as an important force in proto-human evolution was probably a very long one. We must ask, therefore, how natural selection would have operated in the evolution of the secondary motivational system and the emotions through which it was expressed. In order to investigate this it is necessary to consider another important aspect of evolution in general and mammalian evolution in particular – the evolution of sexually differentiated behaviour-patterns within species.

The function of sex within the evolutionary context – that is to say, the reason it confers a selective advantage upon those species which practise it – is the constant shuffling of the genetic deck of cards. Thanks to sexual reproduction a gene pool is kept well and truly stirred and does not 'stagnate'.

Sexual differentiation in terms of physical form occurs in many creatures scattered throughout the whole spectrum of the life-system. The basic pattern of differentiation is between sperm-cell and egg-cell; the former is generally streamlined as a mobile gene-carrier, while the latter tends not to be mobile but to carry the food-reserves necessary to support a developing embryo. In many primitive species the sperm is a luxury, not actually necessary to reproduction, and the male – the kind of body-machine designed by the egg to make and deliver sperm – is really just an added extra. A great many species economise by combining male and female into single individuals, and some make the male a parasite feeding on the female.

Throughout the life-system there is no doubt that the egg is the more important of the two germ-cells. In many arthropod species the male, having outlived his usefulness as soon as the sperm are delivered, is recycled to help provide for the egg as the female eats him after or during coition. In most advanced species, where egg-production is sophisticated and more economical, the wastage rate of sperm remains tremendous − millions are produced for each egg fertilised.

There is no space here to discuss the selective advantages of the many kinds of sexual differentiation exhibited by various orders of invertebrates. Our concern is with the evolutionary advantages conferred by such manifestations of sexual behaviour as are found in mammals and birds, and most particularly with behavioural adaptations.

The parental behaviour of birds is relatively simple, and might be carried out by either male or female. There are, in fact, many species where both male and female sit on the eggs and search for food to supply the fledglings. In other species, however, we see both physical and behavioural sexual differentiation. Males are often brightly coloured while females are brown. Males compete with one another territorially and may play little or no part in sitting on eggs or feeding and protecting the young. This is not merely a form of specialisation − a division of parental labour − for in many instances it is difficult to see how the male's behaviour-patterns can in any way work to the benefit of the young.

It is clear that in these instances, selection operating on the male of the species and selection operating on the female have operated in different ways. Different aspects of behaviour have been favoured by selection, and this reflects the fact that there are different evolutionary priorities operating in the design of efficient male body-machines. These different priorities are not connected with subtractive selection but with maximisation of reproductive effect.

The successful bird is the bird with the most successful offspring. With respect to the female this means that the success of her genes is determined by the success she has in laying eggs and rearing fledglings. The male, on the other hand, can achieve genetic success only through a female. He cannot lay his own eggs. This means that when the degree of parental care required by the offspring is such that it can adequately be carried out by one parent the male becomes − from the viewpoint of the female − a mere convenience. His contribution to the success of the offspring may be made in a matter of minutes. Only when the conscription of two parents to full-time parental behaviour makes a significant difference to the probability of the offspring's survival and success is there powerful selection operating to duplicate the female's parental behaviour in the male. In the absence of such selection other selective factors become more important, and pressure forces differential selection of behaviour-patterns. But what are these 'other selective factors'?

Basically, it is a matter of mate-choice. The bird with the best chance of achieving reproductive success through its offspring is the bird which chooses the mate with the best genes. The priority of the male is always to select the female which will be most effective in parental care. In some instances, the priority of the female is the same, but in others − where parental behaviour in the male is not advantageous − the priority of the female is to select the male who is most efficient in looking after himself, the 'fittest' male with the 'best' genes. This introduces a new measure of subtractive selection into the situation − the males which achieve genetic success are those best-equipped to attract a mate. The methods employed by male birds to attract mates are quite simply those which work − establishing territory, singing most loudly, looking most handsome. Whatever enables a male to be selected at the expense of other males is subject to very strong selection. Those characters which succeed in the long run will be those carried by the males which actually are physically 'fittest' − the reproductive success of accidentally effective characters will be offset by the reduced efficiency of the offspring.

It is this kind of selection, caused by different evolutionary priorities acting in male and female, which is responsible for the often complex and sometimes highly bizarre plumage and mating behaviour of many bird species. The principal selective force acting in the female is directed towards better parental care, while that in the male is directed towards securing a female to bear the young.

In some birds and many mammals, this discrepancy in selective pressures acting in the two sexes is pushed towards its logical conclusion as the male criterion of success becomes not whether he can attract a female, but how many he can attract. In many mammals the individuals form social groups in which one dominant male has a 'harem' of females, and other males may only achieve eventual reproductive success by becoming the dominant male. This is generally achieved by determining status through a 'pecking order'. Dominant males confirm their status by fighting but not by killing – beaten males accept their subservient position on a temporary basis. This kind of elementary 'social system' reflects the fact that mammals have, on the whole, much more complex behaviour-patterns than birds. Their brains are large enough to permit natural selection much more leeway in sculpting behaviour-patterns appropriate to the priorities operative in the two sexes.

The greater versatility of the mammal brain permits greater differentiation between behaviour in the sexes in some respects, but in others it tends to reduce the differences. Parental behaviour in mammals is more complex, more efficient, and of longer duration than in birds, and the male is necessarily co-opted back into the business of child-care and child-protection. The male's co-operation can once again make a very significant difference to the probability of the offspring's success. And it is in the mammals that we see what does not tend to happen in birds – the specialisation of parents in different aspects of parental behaviour. The male often becomes important as the protector of both child and female – in many mammalian species the male is larger and more powerful than the female. This applies especially to species where the harem system operates, because the establishment of the pecking order facilitates strong selection between males with regard to physical strength.

In mammals, therefore, protection of children often becomes a two-tier system. Individual females look after their own offspring while the dominant male provides protection for the whole group. In behaviourally sophisticated species the dominant male is generally assisted by the subservient males, whose own chances of genetic success are tied up with the future of the group. Their only hope of eventually reproducing their own genes is that at some time in the future they might inherit the harem, and therefore it is in their interests to add their own strength to protecting it. In this way mutual co-operation between competitors becomes selectively favoured under certain circumstances.

Some situation similar to this probably applied in the anthropoid apes ancestral to man. It is within this kind of a context that we may hypothesise the initial complication of the motivational system to include primitive emotions. As I have already said, the most useful emotion in this context would be love – the mother's love for the child which increases the priority of child-protection. Such an emotion would not be so advantageous in the male.

Increasing brain-power, however, must necessarily introduce a strain into the elementary social system as the strength of the headman, by which he maintains his position, becomes matched by cunning in the subordinate males. The selective pressure acting on subordinate males within such a group is very strong. Any reproductive success gained by a subservient male is very significant indeed, and any factor permitting him to pass on his genes would automatically be successful within the gene pool of the species.

There are many behaviour-patterns which enable subservient males to increase their chances relative to those of the dominant male. Behaviour-patterns which lead the subservient males to band together to overcome the headman's superior strength would be favoured (although the subsequent battle over determining how the harem is to be redistributed afterwards would introduce an element of chance, reducing the selective pressure). Similarly, ways of fighting in which the possession of simple strength is devalued are also favoured by selection, so that subservient males would be under selective pressure to create weapons. For convenience, we may call these selective factors the *revolution syndrome* and the *David-and-Goliath syndrome*. They are operative only in male behaviour-patterns and limited in their effect, but they cannot be ignored. (Ancient history testifies eloquently to the difficulties involved in maintaining a headman system in human societies – 'uneasy lies the head that wears the crown'.) There is, however, a far more effective and most advantageous behavioural strategy which would be subject to strong selective pressure in the subservient males. This is known as the 'cuckoo principle'. The cuckoo, of course, is a bird which exploits the parental behaviour-patterns of other birds by laying its eggs in their nests. Its offspring get the best of parental care while the cuckoo does nothing. Most of the cuckoo's victims are species in which both parents collaborate in the feeding of the young, because the young cuckoo is very voracious. The cuckoo is an isolated case – rather an oddity – because this strategy, in birds, can only be used with the utmost discretion. Too many cuckoos would eliminate the host species – the continued reproductive success of the cuckoo is involved with a strict statute of limitations.

Birds, of course, have rather limited brain-power. The species parasitised by the cuckoo cannot identify the intruder. The reason that the cuckoo strategy is not more widely used among the higher vertebrates is probably the fact that most of them can identify their own offspring and will not tolerate a usurper. The cuckoo principle is thus not apt to be favoured by selection in an interspecific context. Within a single-species group, however, things are different.

A female, of course, usually knows her own children – she, after all, gives birth to them. But within a group containing several males, paternity is much more difficult to establish. The protective behaviour of the dominant male is geared to the expectation that he is securing the future of his own genes. If, however, a subservient male can impregnate a female, he can exploit the protective social structure of the group for his own ends. This is much more efficient, selectively speaking, than managing revolutions or killing the dominant male by superior skill with weapons, for it exploits the strength of the dominant male as well as his weaknesses. Logically, natural selection within proto-human groups must favour the cuckoo principle to a far greater extent than the revolution syndrome or the David-and-Goliath syndrome, although all three would have some degree of survival value.

The most important forces causing change in behaviour-patterns in proto-human society might therefore be assumed to be the evolution of love (in females) and the evolution of cuckoo-strategies (in males).

As brain-power increases, the tendency for proto-human groups to operate on a rigid headman-system like that which applies, for instance, to the baboon troop, must decline. As methods of subverting the system become incorporated into behaviour-patterns, the selective pressure maintaining it dwindles. The end result is bound to be the limited redistribution of the harem among the males of the group. In order that the strength of selection maintains behaviour-patterns in the males which contribute to the care and protection of the young, some system must evolve by which the males can identify their own offspring just as the females can. There will thus be a tendency to adopt a new system of male/female association. While not necessarily monogamous, practicality will ensure that harems are limited in size and confined to the most dominant males (and dominance, at this point, begins to imply factors other than simple physical strength).

At this stage we are still dealing with behaviour-patterns which are, to a very large extent, genetically programmed – i.e. instinctive. But there is already in the population a second system of motivational organisation which may be co-opted into the organisation of efficient social groups – the emotions. Analogues of maternal love in males can, at this stage, become selectively useful, and the emotional spectrum will begin to differentiate to adapt to the circumstances. The romantic/sexual love of the male for the female will now become a selection-favoured character – and so will the sexual jealousy of one male for others.

Once the basic facility for the father to identify his children is provided by social structure the possibilities for the complication of the emotional capacity of the individual is increased greatly. The co-opting of other evolutionary priorities into the emotional system also becomes possible and selection-favoured – perhaps the most important emotion here is pleasure, which is a feeling connected with the more basic motivational system involved in behaviour-programming. Broadly speaking, the things we find pleasurable are the things which we need to do in order to facilitate genetic survival.

It is important to realise that as emotion-based behaviour increases instinct-based behaviour must decline. Emotion appears initially as a system by which programmed behaviour can be modified, but ultimately – as the emotional system evolves and becomes complicated – instinctive reactions are very largely replaced. This process of replacement puts an ever-heavier burden on parental guidance, as the amount of behaviour the infant has to be taught rather than being equipped with at birth increases vastly. Similarly, the learning capacity of the infant must increase. Neoteny becomes a vital factor in proto-human evolution.

The switch from genetically programmed behaviour to emotionally governed behaviour is only half the story, however. The replacement of the one by the other paves the way for a new system of choice-making much more efficient than the emotional system: rationality. The original functions of rational creative thought within the evolutionary context must be sought in terms of the particular strategies which were selectively advantageous in proto-human society. Once the capacity to reason had evolved, however, it was, like the capacity for emotion, amenable to conscription in the service of many different evolutionary priorities.

While this was happening, however, something much more important was also happening. This evolutionary sequence is favoured by natural selection because choice, *per se*, is a selectively favoured character. Versatility is worth a great deal in the 'struggle for existence'. Natural selection favours the capacity for choice – it does not dictate which choices shall be taken. The ultimate product of natural selection is a species which is no longer subject to it, a species with the intellectual equipment necessary to opt out of a growth-trap, if it can perceive one, to oppose the natural forces of change.

Gradually, therefore, as man became an emotional creature, and eventually a rational one, the selective forces shaping his nature and his society disappeared, to be replaced entirely by the forces generated within the conscious mind – the priorities of emotional choice and rational decision.

In human nature as it exists today, and as it has been reflected through history, we can undoubtedly identify the echoes of the selective priorities extant in proto-human society. We can identify the headman principle and the syndromes opposing it. We can identify the cuckoo principle and its insidious influence on human behaviour. We can identify differences in the many kinds of love we encounter, which can be correlated with selective forces operating during the differentiation of the emotional system under natural selection. We can, in short, find a good deal of 'human nature' in a logical study of the evolution of man from his non-human ancestors. How could it be otherwise?

But we must not forget that we now have the intellectual power to control and overrule the aspects of human nature inherited from the proto-human group. We may, by reference to the struggle for existence and the survival of the fittest, explain human nature, but we can in no way excuse it. It has been argued by some writers, who have attempted to rationalise human behaviour in terms of analogies with animal behaviour, that human beings should recognise the 'failings' of human nature and make accommodation for them in the way they organise their society. They argue that human nature cannot be overcome. But this is nonsense. The whole usefulness of human nature, in the evolutionary context, is that it is flexible and changeable – amenable to modification by choice. Human nature evolved in order to be whatever we can make it.

This, then, is the story of biological evolution – how it acts and the sort of thing it produces. But what sort of process is cultural evolution – the process which has succeeded it?

There are two aspects of evolution in human society to be considered. The first – and less important – is the analogue of natural selection by contemplation of which Darwin first discovered his theory: eugenic selection.

The crops we grow and the animals we have domesticated are all products of eugenic selection. What this entails is taking over the role of chance in determining which animals or plants shall contribute most to the next generation, the strategic management of gene pools. The essential difference between natural selection and eugenic selection is that where nature selects the most efficient forms, eugenic managers select the most convenient. Thus, when a New England farmer in the eighteenth century discovered among his crop of lambs a mutant with short, bowed legs, he used it to found a whole new breed of sheep – the Ancon sheep – which, though they were no more efficient at producing wool, mutton or more sheep, were very convenient because they could not jump over fences.

Perhaps the most eloquent testimony to the potential changes which can be introduced into species in a relatively short time (short, that is, on the evolutionary time-scale – no more than a few centuries) is the wide range of physical forms produced by dog-breeders. Genetically, a Pekinese and a Great Dane are not so different that they must be reckoned separate species, but they are very dissimilar in their external characteristics. Without the aid of growth-traps, dog-breeders have managed to push the species towards several of its mechanical limits, and many breeds are handicapped by deliberately cultivated physical disabilities.

The use of eugenic selection in human populations to 'improve the human race' has been suggested ever since Darwin – his cousin, Francis Galton, was one of its earliest proponents – although Hitler's Germany has been the only nation ever to attempt to institute such a programme. Galton's book *Hereditary Genius* caused something of a stir in 1869, and a pamphlet called *Daedalus* written by J. B. S. Haldane in the 1920s, advocating human genetic control, similarly provoked a good deal of reaction (Aldous Huxley's *Brave New World* is a condemnation of the future as envisaged by *Daedalus*). The controversy begun by Galton as to whether great men are born or made – created by 'nature' or by 'nurture', as he himself put it – still goes on.

In laboratory animals, of course, behaviour and temperament

are far more influenced by genes than by rearing, but the whole of human evolution has been concerned with the reversing of that balance. The nature of the human animal is such that far more can be gained in terms of improving individuals or groups by improving the conditions of their upbringing than could possibly be gained by centuries of the most rigorous eugenic selection. Even if that were not so there are other considerations to be borne in mind. Eugenic selection, as I have said, is selection for convenience. The question of who is empowered to choose the direction and the manner of eugenic selection thus becomes most important.

Modern exponents of eugenic selection for the human race generally espouse 'positive eugenics' rather than 'negative eugenics' – that is to say, they do not want to sterilise the 'unfit' but merely wish to encourage the most desirable specimens of humanity (whoever they may be) to breed more. While it may not be as cruel, this policy is probably just as futile, as larger families imply less efficient upbringing.

There has always been a measure of unnatural selection operating in human society. Artificial increases and decreases in the survival-probability of individuals in certain human groups has been manifest in a number of ways. Until the twentieth century war was far more likely to kill young healthy males than other members of the population, although the mechanisation of war has made certain that women and children suffer too. Owing to a wide disparity in sanitary conditions, the plagues of recent centuries always carried off a far greater proportion of the poor than the rich. All this, however, is hardly eugenics.

And yet, without the aid of eugenic selection, and in a space of time too short for natural selection to have had any material effect, there has been considerable human evolution – physical form has altered relatively little, but ways of life have altered tremendously. These changes are due to the second evolutionary force important in the human species, which is not in any sense selective.

The primary philosophical opposition to Darwinism has always been provided by some variant of Lamarckism – the doctrine that acquired characteristics can be inherited. Acquired characteristics, in this context, are not genetic mutations but changes brought about during the lifetime of an organism extraneous to the influence of the genes. A standard anecdotal refutation of Lamarckism is the observation that though the Jews have been circumcising their male offspring for thousands of years, nature has not accommodated them by producing infants who conform to the custom without the necessity of surgery.

So far as we can ascertain, Lamarckian evolution does not occur in nature. The Russian agronomist Lysenko, who attacked Darwinism in the name of communism, and the American flower-breeder Luther Burbank, who tried to discredit it in the name of fundamentalism, both enjoyed a degree of political success, but were both eventually discredited themselves.

The fact that Lamarck was wrong in suggesting that giraffes grew long necks because generations of their ancestors kept stretching what necks they had to reach leaves which grew higher up on the trees should not, however, blind us to the fact that cultural evolution in the human species is definitely and completely Lamarckian in kind.

The Lamarckian model of evolution failed because there proved to be no natural mechanism by which an animal could communicate to the next generation characteristics it had acquired and found useful in its own life. This is true when applied to physical evolution but *not* when applied to behavioural evolution. In the mammals, and to a lesser extent in the birds, Lamarckian evolution becomes a possibility. In the human species such mechanisms *do* exist, and they work very well indeed.

The basic component of the Lamarckian mechanism is language. As employed by other mammals and birds, language seems to be a very limited tool and the extent to which animal evolution has been influenced by Lamarckian forces is probably negligible. The information-carrying capacity of animal languages is not great (one interesting exception may be the languages of the social insects, whose position in the invertebrate hierarchy is analogous to man's status among the vertebrates) and for the most part only simple messages are exchanged; warning signals, identification signals and challenge signals are the most common. These systems of information-transmission are far less complex than the genetic systems of the simplest living organisms. The human language, however, has evolved towards a similar level of complexity.

The organisation of the gene-system is remarkably similar to the organisation of some human languages. A sequence of three bases codes for a single amino acid. As there are four bases there are $4 \times 4 \times 4$ possible 'codons', but in actual fact there is a certain degree of degeneracy – the number of different 'meanings' conveyed by the codons is just over twenty, so that in some cases different codons have the same meaning. The 'genetic language' thus consists of four 'letters' which may be arranged into 'words' of three letters each. Every protein produced by an organism is prescribed by a 'sentence' which may contain anywhere between eight and several hundred words. Each chromosome is a 'book' containing many sentences strung together although there is no 'narrative sequence'.

It is instantly obvious that human languages have much more information-carrying capacity than the genetic code. We have twenty-odd letters instead of four. There is a very considerable degree both of degeneracy (i.e. words with similar meaning) and of redundancy (i.e. words without meaning) in the manner of their organisation into words, but the number of letters in a word may vary from one to ten, or even more. The number of meanings which can be conveyed is thus very large. Only when it comes to arranging the words into sentences are human languages more economical than the genetic one – but then, they can afford to be.

The message carried by the genetic code had to be put together by trial and error. Messages which worked were perpetuated, those which did not were rejected. Not only is this a slow process and a wasteful one, but there is no capacity for strategic re-coding: there is no way for an organism deliberately to rearrange or change its own genetic heritage. But human languages are more versatile. New words can be invented to fit specific meanings, old words can be re-adapted to change their meaning. A man with something to say does not have to wait for chance to throw up the words with which to say it. Human language is modified by choice and not by chance.

Language provides a means by which a man can transmit all his acquired characteristics to his children. An animal which learns something new can only give to its offspring the genetic potential for making the discovery – which *might* allow it to be made again. In the human species there is no such difficulty – through the use of language man has mastered Lamarckian evolution.

The simplest linguistic process of all – that of *naming* things – is a means of abstracting information from the environment, encoding it, and passing it on. Sometimes one name describes several things, or the same thing may be known by several names. Sometimes names are similar in sound and form, and some names are expressive of feelings and reactions. All these are instances of strategic coding, so that a new generation, in the process of learning what the world contains, also inherits the discoveries and judgments of previous ones.

Out of language grows myth – a system of encoding ideas. Myths are the earliest thought experiments, and though we tend nowadays to think of myths as ancient stories, we have not lost either the need for or the use of them. While we turn old myths into stories we create new ones to take their place. Modern mythology is contained in advertising, in popular literature and in the news media.

Writing is an advance in the method of communicating information rather than an advance in coding capacity itself, but writing permitted the ordering of knowledge and the perception of structure within it. It is not a coincidence that those cultures which first discovered scientific knowledge and made significant advances in science were the cultures which made most use of writing, and that those cultures which remained scientifically naive were those which did not. (Similarly, it is significant that the post-Renaissance revolution in science in Europe followed closely the invention of the printing press.)

It was writing which allowed knowlege to become a thing in its own right, divorced from the men in whose minds it had hitherto been confined. Knowledge handed on by word-of-mouth, in myth and custom and law, remains always mutable and always changing. Writing permitted, for the first time, the concept of immutable knowledge – the idea of permanent, unchanging principles so essential not only to science but to religion. Before there was writing there was religious belief, but the kind of institutionalised religion we know today is usually based on 'sacred writings' – beliefs made permanent by their isolation from the believers. (Again, printing – which took the holy word out of the hands of monks and made it available to the people – came shortly before the wars of religion and the Reformation.)

Language is not the only means by which acquired characteristics can be inherited. Another is wealth. Man became a tool-using animal, and the tools he made and used could themselves be passed on to his children as well as information about how to make, and use them. Natural selection in proto-human groups would have favoured a kind of 'guild system' whereby men passed on their knowledge and their tools to their own children and them alone – and it would also have favoured thievery as a means of acquiring property other than by creative effort. If theft and filial inheritance persist today, however, it is through choice and not through chance. Alternative social organisations have existed and do exist in which wealth is held in common and theft is rare.

Language and wealth are the mechanisms of Lamarckian evolution in human society. By the use that various social groups have made of these means we have made of ourselves what we are. It has been argued that to say 'man invented tools' is to employ a misleading perspective, and that it would be more accurate to say that 'tools invented man'. In actual fact neither perspective can stand alone. Man evolved through a feedback process in which he was changed by his own creativity. Man invented tools *and* tools invented man – and this is not so much a circular argument as a spiral one. Man and his tools have changed in phase with one another, but they have never ceased to change. This process has brought us the power of conscious, rational thought, and with that power we are armed to seize control of human evolution, not by eugenic selection, but through the Lamarckian processes. With these instruments, we can restore to the pattern of evolution on Earth that which Darwin demonstrated as non-existent in nature: an evolutionary goal. In fact, if we realise the true nature of human evolution we have something which even the ideological opponents of Darwinism do not profess – a *choice* of goals. We are not committed to the particular destiny which philosophers once sought to find built in (by God) to the pattern of evolution.

On the other hand, of course, we no longer have that guarantee ...

9 The Seat of the Soul

Descartes believed that the 'driving seat' of the human body-machine was the pineal body, a small organ at the base of the hind-brain which appears to have no particular function in man. (In some lizards – notably the tuatara – the pineal body appears to be the vestige of a third eye situated in the back of the skull.)

It is worth noting that the answer to the question which prompted Descartes to make this inaccurate guess is not so obvious as it might seem. The question, basically, is 'Where am I?' One classic reply is 'In my skin', but this is really too vague. Certainly, everything within my skin is me, but I can lose a substantial portion of it without the essential me – the self-conscious me – becoming depleted. If I had my leg amputated I would not automatically lose memory or intellect or any other mental faculty. One can conduct a rather gruesome thought experiment in which every bit of me that can be removed without impairing the essential me *is* removed – and what is left becomes the answer: that is where I am. This way, all my limbs and most of my organs can be removed, the functions of my heart and lungs can be undertaken (temporarily, at least) by a machine. Virtually all that is left is inside my skull and my spine. Am I, then, in my brain?

Already there are large numbers of people who would disagree with me. Many human societies, of the kind we sometimes call 'primitive', would find the idea that the essential identity of a man is located in the brain and spinal column quite alien. Many of them believe the seat of the soul to be in the abdomen – that is where they generally feel strange if something is wrong with them. But let us, for the time being, leave this opinion aside and continue with our own line of thinking. I am in my brain, but where, exactly?

During the Second World War, a Russian army officer named Lev Zassetsky had part of his head shot away. His personality was relatively unaffected, but his ability to perform many mental functions − reading and remembering among them − was impaired. Over a number of years, however, he managed to retrain what was left of his brain to read, and he recovered his memory to a considerable degree. Then again, there was a fashion not so long ago in psychiatric medicine for pre-frontal lobotomy, in which disturbed patients had large chunks of their forebrain removed because of supposed therapeutic effects. This practice tended to turn a lot of the patients into human vegetables, but in some cases anxiety was relieved without a great deal of memory-loss and impairment of other mental faculties. As long ago as 1808 Franz Josef Gall, founder of the failed science of phrenology, had demonstrated that different parts of the brain were concerned with different functions within the organism, but it proved very difficult to map these functions − most especially, it proved almost impossible to detect which area of the brain contained the self and its attributes: the seat of the soul.

Eventually, it was concluded that the self was not located in a specific part of the brain, but was distributed throughout it. It was a product of the whole rather than a part or a sum of parts. The self is not an organ, like the pineal body or a lobe of the brain, but a state of organisation. It is not, itself, material but a pattern within matter.

This distinction is vital. It means, for instance, that Descartes, who was almost right 'geographically speaking' in putting the soul beneath the brain, was, in a way, far more wrong than the primitive peoples who place the self within the entrails. They, it seems, may have a clearer idea of what kind of thing the self/soul really is, a diffuse thing which is unconfined by the structure of bodily matter.

In a sense, the failure of anatomists to locate the self specifically within the brain presents a challenge to the kind of logic we used in narrowing down our field of search to the brain in the first place. The fact that the self survives while bit after bit of the body is removed does *not* prove that the self is not 'in' those bits. It merely illustrates the extent to which the self may sustain itself against the erosion of the material structure on which it is superimposed.

There is an anecdote about an old lady whose transistor radio had gone dead, and who was advised by her small grandson to take out the battery and stand it upside-down on the mantelpiece for a while, so that all the electricity left in the battery would flow down to the electrodes. This advice is based on a false analogy, as if electricity were contained by a battery in the same way that toothpaste is contained by a tube. In reality, the physical organisation of the matter inside the battery is such that electric current may be generated – until the battery 'dies'. In the same way, it is a mistake to think – as Descartes did – that the brain 'contains' the self (or the soul). The self is a *property* of the body-structure, which, though it can be retained by the brain, is something which really involves the whole body.

Science came very late to the study of the self. Despite Descartes' conviction that bodies were machines and thus belonged to the realm of scientific inquiry, the attitudes which grew out of the Cartesian partition condemned scientific medicine as profane tampering with God's handiwork. Because of this the advance of medical science was drastically handicapped, and its practitioners, like the exponents of Darwinism, were forced to extremes in the way they looked upon their own endeavours. They became biological mechanicians, involved in rationalist studies, quite unconcerned about those aspects of the human being which could not be revealed by dissection. Just as the polarisation of ideas robbed the Darwinian theorists of the intellectual middle-ground which they needed to put their ideas into perspective, so it prevented many theorists in psychology from adopting the perspectives which might have allowed them to begin the task of making sense of human behaviour and character in the late eighteenth and early nineteenth centuries. Gall's phrenology was by no means as silly as it seems today – it was merely an attempt to approach the problem from the wrong direction, a direction of approach quite natural considering the intellectual climate of the day.

The prospectus for a genuine science of psychology was first explored in detail by Herbart in *Psychology as Science*, published in 1824–5. He pointed out the limitations forced upon such a science by the inability to experiment with the mind, but suggested that this did not render the scientific approach to the problem completely impotent. (As I pointed out in a previous chapter, we cannot experiment with the stars, but this hardly implies that there cannot be a science of astronomy.) Herbart championed the usefulness of mathematics in handling psychological data – an excellent idea, but one which fell apart in his hands because he never quite came to grips with the fact that methods of mathematical analysis can only work in the context of the most rigorously systematic observation. E. G. Boring, a historian of science, lamented that Herbart 'exhibited the not uncommon case in science, in which inadequate data are treated with elaborate mathematics, the precision of which creates the illusion that the original data are as exact as the method of treatment.' (As Boring pointed out, this tendency is one of the principal diseases of modern science and pseudo-science. The false sophistication of statistical methods applied to poor data has led to such well-known axioms as Churchill's comment that 'there are lies, damned lies and statistics'.)

Though Herbart really failed to achieve anything positive in furthering scientific psychology his was an important influence in that he recognised the sterility of other avenues of approach. He had no confidence whatsoever in physiological psychology – the attempt to explain the mind in terms of the brain – because he believed that contemporary knowledge of physiology was woefully inadequate to that task. Though his programme for observational psychology was inadequate, he was, at least, pointing out the direction in which advances might be made.

The three areas of psychological inquiry which co-existed in the early nineteenth century were not so much different branches of a single science as wholly different trees. It was not until the middle part of the century that the three approaches – from philosophy, from physiology and from observation of the mind's activity in human behaviour – began to be integrated into a new science. Thomas Brown, John Stuart Mill and others began to bring some semblance of organisation to the knowledge of how the memory operated and of the correlation between ideas and sense-impressions.

Herbert Spencer provided psychology with the thoroughly rationalistic perspective which he attempted to apply to various other human sciences. He identified instinct as 'compound reflex action', and was completely committed to the notion that the key to psychology lay in physiology. It was Spencer's attitude that Haeckel echoed in *The Riddle of the Universe* when he discounted free will as an illusion. A little later Lloyd Morgan began his pioneering studies of animal behaviour, attempting to begin with the simpler task of unveiling secrets of the animal mind before going on to the far more complex problems of the human one.

Morgan brought to psychology a new experimental perspective – it was true, as Herbart had pointed out, that one could not experiment with the mind, but one could experiment with behaviour. Morgan proposed the famous 'principle of parsimony' – an Occam's razor for psychologists – which stated that behaviour must always be explained by the simplest mental processes capable of accounting for the facts. This principle was useful in opposing the anthropocentric interpretations popularly applied to the behaviour of domestic animals. Like Occam's razor, however, the principle of parsimony could only point out the simplest answer, and the conviction that the simplest answer was always the right one was a matter of faith – a faith which dominated nineteenth-century thinking but which has since been drastically eroded.

The attempt to co-opt psychology into the great pattern of scientific certainty which seemed to be emerging in the nineteenth century prompted a great rash of new textbooks on the subject produced throughout the 1880s and 1890s. Perhaps the best of them was William James's *Principles of Psychology*, which was far more successful than Herbart's work in exposing the limitations of various approaches to the subject. James was far more interested in the study of individual people and the differences between them than in the attempt to expose 'fundamental laws' or the 'elementary qualities of the mind', which was the supposed goal of the psychologist in the eyes of Spencer and Morgan. He was the first to realise the importance of 'abnormal' mental phenomena in helping to understand the way the mind worked. He remained, however, sufficiently dedicated to Occam's razor to oppose the concept of the 'subconscious mind' as a retrograde step in thinking comparable to that of the physicists who wanted to introduce new hypothetical entities like electrons into a science which had only just been swept clean of such mumbo-jumbo.

Ultimately, the breakthrough in psychology came about through the process of systematic observation of thousands of cases of 'aberrant' mentality. Freud collected sufficient data to allow him to formulate a theory of the structure of the mind. The unconscious mind became a necessary hypothesis. Having thus divided the mind into two, he went on to identify three structures of organised motivations: the *id*, the *ego* and the *superego*. This model allowed him to rationalise the psychological changes which take place in the developing mind, and to provide an account of the irrational component of everyday psychology.

It must be stressed that Freud's model of the mind, like Bohr's model of the atom, is only an analogy – an assembly of concepts which help us to understand the kind of forces which are at work in the mind. Like Bohr, Freud did not discover anything – he invented a system of ideas. Just as Bohr's model of the atom had to be progressively complicated and modified to account for more and more anomalies in the scheme of things it was intended to systematise, so Freud's psychology has been continually overhauled and modified, by Freud himself and by those who came after him.

The chief difference between models of the atom and models of the mind is *not* the kind of entity that they are trying to explain. Neither the atom nor the mind is a solid, material thing, but rather an assembly of phenomena. In each case the evidence to which we refer in order to check the 'rightness' or 'wrongness' of the model is a secondary effect of the organising principles we are trying to codify. In the case of the atom the checkpoint to which the model has to be compared is the hydrogen spectrum, with its inbuilt mathematical properties like Balmer's Ladder. In the case of the mind, however, the reality is something which is far less easily qualified, let alone quantified. We have no elementary 'hydrogen mind' which exhibits specific and quantifiable behaviour, despite the attempts to force the laboratory rat into such a role. We have, instead, the whole elaborate spectrum of human behaviour.

It has been argued that Freudian psychology is so vague that it can be perverted to provide an explanation for absolutely any set of facts imaginable – that the limits of its applicability are untestable. This is true – but it is the ubiquity of the data rather than the ubiquity of the theory which is to blame. Psychology cannot go the same way as atomic physics, with its concepts becoming progressively more mathematical as they are subjected to the cunning discipline of ever more complex equations. The nature of the frame of reference prevents that. There can be no psychological equations, no gentle guidance from mathematical aesthetics. There will never be a Heisenberg or a Dirac to make the Freudian model of the mind complete and fully comprehensive.

But if we are never going to get a description of the mind as accurate and sophisticated as the description we have of the atom, what kind of description can we get? This is a question which has raised a great deal of controversy in twentieth-century psychology. There have been two main directions of development, the first of which we may illustrate by considering the ideas of Carl Jung, and the second of which is associated with a school of psychology called behaviourism.

Jung developed Freud's model of the mind along lines which completely ignored the principle of Occam's razor. Freud had seen the unconscious mind as a purely personal thing, containing experiences and memories which had been 'repressed' by the conscious mind – the debris of the battle between the id, a motivational complex, operating exclusively on 'the pleasure principle', and the superego, a kind of 'moral censor' imposed upon the ego (the self-conscious, decision-making part of the mind) by social conditioning. He had found all these entities necessary to an understanding of human psychology, and no more. As his work progressed he changed his mind somewhat concerning exactly what these factors were and how they operated, but his basic concepts remained more or less unaltered. One of his contemporaries, Adler, challenged his idea of the supremacy of the pleasure principle and invented the inferiority complex instead, but left the basic pattern unchanged. Jung, however, felt that the model was not complex enough, and that a more elaborate structure had to be hypothesised.

Jung felt that the unconscious mind was not simply a collection of elements repressed from the conscious mind. He termed Freud's unconscious the 'personal unconscious' and added to it a 'collective unconscious' whose contents are similar in all of us, being genetically determined, and whose influence upon our minds is thus common to us all. The contents of the collective unconscious, according to Jung, are the *archetypes*: entities which are expressed in dreams and in conscious thinking as archetypal symbols – symbols which we all recognise and to which we all, to a large extent, attribute the same meanings. The symbols which derive from the archetypes crop up in myths and therefore, according to Jung, provide the elements of the way that human beings attempt to find order in the world and regulate their behaviour within it. Among these archetypal images are the Great Mother, the Wise Old Man and the God-image (which is derived from the archetype of the self).

Jung's psychology is Freud's re-mystified. There could be nothing more contrary to the spirit of nineteenth-century science, and perhaps of science itself. Jung's psychology is more like a mythology than an exposition of mechanistic laws and principles by which singular phenomena can be integrated into a whole system of understanding. But the twentieth century has seen the re-mystification of the physical sciences made necessary by new discoveries, and we must ask whether the mystification of psychology is necessitated by the nature of the data with which the science works.

The behaviourist psychologists were adamant that such re-mystification was not necessary and was, in fact, highly undesirable. The school of thought was founded in the 1920s by John B. Watson, and drew its inspiration primarily from Lloyd Morgan's principle of parsimony, and from the work of I. P. Pavlov.

Pàvlov's dogs remain in common parlance as the classic example of the *conditioned reflex*. When the dogs were fed, a bell was rung until, after many repetitions, whenever the bell was rung the dogs would salivate. The bell and the food had become 'associated' in the mind of the dog, so that the former stimulus prompted the reaction appropriate to the latter.

According to the principle of parsimony, all behaviour ought to be explained with reference to the simplest mental process, and the conditioned reflex was undoubtedly the simplest mental process amenable to use in explaining behaviour. The behaviourists came to believe – and set out to try to prove – that the conditioned reflex was the basis of all behaviour, animal and human, and that the mind had no other attributes. This was a bold decision, but one well in accordance with Occam's razor. The conditioned reflex could be demonstrated by experiment, and was therefore scientific. All other mental processes could not be so demonstrated, and therefore science had no business with them – they were all illusions and imaginary concepts.

According to the behaviourists, psychology consisted of the study of stimulus and response. Stimuli had to be quantified and responses had to be described exactly and without any interpretative language. One could not speak of a response as 'fear' or 'pleasure', one could only describe exactly what the responding subject did. Anything that might happen in between stimulus and response was unobservable, and hence irrelevant.

The methods used by behaviourist psychologists are famous for their devastating simplicity. Their experiments generally involved putting rats in boxes and training them to press levers, stimulating the rats to adopt particular response-patterns with the judicious use of food pellets (to 'reinforce' behaviour-patterns by reward-conditioning) and electric shocks (to destroy potential behaviour-patterns by avoidance-conditioning).

By this means the behaviourists decided that they could completely rationalise the behaviour of the laboratory rat by assuming a system of three 'primal drives' (hunger, thirst and sex) to which were linked instinctual behaviour-patterns modifiable by conditioned reflex. The rat could therefore learn – its behaviour was adaptable – via the association of stimuli and reinforcement-by-reward or avoidance-through-pain. Many behaviourists became extraordinarily adept at training young animals to perform quite complex behaviour-patterns.

They ran into some difficulties in trying to extend their theories to account for human behaviour – responses to stimuli tended to be so much more complicated and difficult to describe. They had to reintroduce the concepts of 'fear' and 'pleasure' to describe human reactions, although they tried to make it perfectly clear that what they were describing were not feelings but combinations of observable and describable behavioural characteristics.

As time went by it became progressively clearer that not only could human behaviour not be rationalised by analogy with a rat in a box, but that even a real-life rat was a good deal more accomplished than a behaviourist laboratory rat. Some behaviourists attempted to patch up the model by introducing new drives (the 'exploration drive', for example) and allowing more and more interpretative language into their accounts, but in the end what showed up clearly was the total inadequacy of their perspective to cope with the data.

There is no doubt that the behaviourists were thoroughly scientific. Their approach was rational, and their determination to see what could be done with the observable facts and the observable facts alone was commendable. The simple fact is, however, that they failed. They attempted to discover a rational psychology and found no psychology at all. Curiously, behaviourism survives today – not as a scientific theory, but as a philosophy. Though behaviourism may have failed, there are nevertheless many hard-headed scientists who believe that its approach was the right one. They remain convinced that behaviourism must be true because, damn it all, it's scientific.

The legacy of behaviourism has left in the minds of laymen and many psychologists an altogether disproportionate idea of the role which conditioned reflexes play in human behaviour. This is not because any psychologist has ever found any convincing evidence for the importance of conditioned reflexes in human behaviour, but simply because it is *the only kind of explanation that many psychologists have ever looked for*. The fact that the behaviourists found no forces active in determining human behaviour other than the conditioned reflex is not surprising when one remembers that they had decided before they started that the only forces they would accept as real were those which could be observed and described by their methods — i.e. conditioned reflexes.

We should, I think, be prepared to realise that the very factors which make the conditioned reflex the most observable of behavioural mechanisms also make it the most modifiable one. The fact that it is practicable and fairly simple to condition a dog to salivate, or to habituate a rat to pressing a lever, does *not* imply that all naturally acquired dog/rat behaviour arises from the same process. H. G. Wells once wrote an article in which he pointed out that Lloyd Morgan's dog, by applying the logic of the principle of parsimony to its master's behaviour, might come up with as low an opinion of human intellect as Lloyd Morgan had of canine intellect. A rather more forceful exposé of the limitations of the scientific method developed from the same line of thinking is presented in Franz Kafka's short story 'The Investigations of a dog', which tells how a rationalist pet dog achieves a perfect understanding of his environment and the natural laws underlying the events taking place within it without having to admit to the existence of such crude hypothetical entities as human beings. A triumph for Occam's razor — and complete self-delusion.

In Part One of this book I have taken pains to point out the limits of what we can know about the microcosm and the macrocosm. In dealing with concepts which do not behave like analogues of the objects we see around us, and about which we can obtain only indirect or incomplete data, there has to be an element of doubt about what we call 'the truth'. Our understanding of the very small and the very large is necessarily limited. The same argument also applies to the mind. The data which we can obtain are indirect and incomplete, and there is the same kind of limitation on what we can know and find out through scientific observation. The mind, like the atom and the universe, is a hypothesis, and — to use de Sitter's words — 'must be allowed to have properties and to do things which would be contradictory and impossible for a fine material structure'.

We must, I think, recognise that Jung's psychology is not

scientific in the sense that behaviourist psychology is. More than any other scientific doctrine it has recovered the legacy of mystical intuition. Jung's conclusions are not, for the most part, amenable to any kind of scientific 'testing'. They stand or fall not by reference to Popper's scientific philosophy of potential falsification but in a purely pragmatic sense – they are as good as they are useful in helping us come to terms with the material of the psyche. But we must also recognise that to be unscientific is not necessarily to stray further from the truth than those who remain devotedly scientific even when their methods have proved quite inadequate to cope with their data. Behaviourism is scientific, but quite incompetent to explain human behaviour. Jung's psychology, though its own competence is open to question, is at least making the attempt that the behaviourists declined to countenance.

While Jung and the behaviourists were taking observational psychology in two entirely different directions during the 1920s, Hans Berger was making a discovery which was ultimately to open up important new ground in the field of physiological psychology. He inserted silver wires into the skin of a man's scalp, at the back of the head and on the forehead, and connected them across the most sensitive galvanometer available to him. He found a very slight but distinct fluctuation in the potential difference between the two electrodes.

It had been known since 1875 that there were electric currents in the brain and that thought processes might therefore be electrical in nature. The discovery of these fluctuations in the electrical activity opened up speculation about 'reading' thought processes in the brain – or at least correlating kinds of mental activity with kinds of electrical activity. Berger's work was treated with a great deal of initial scepticism, and some years were to pass before the development of better galvanometers (aided by the thermionic valve and its use as a current amplifier) allowed unquestionable demonstration of the 'Berger rhythms'.

The Berger rhythms quickly became an interesting scientific puzzle, rather like the hydrogen spectrum. The fluctuations could be mapped by an encephalograph – which, in its most primitive form, was simply a pen attached to the galvanometer needle which drew a trace on a revolving drum. The trace so obtained was, however, a set of squiggles with little apparent pattern. Decoding the trace involved sorting out the various regular cycles which were superimposed upon one another to make the compound curve drawn by the pen. This was not easy. Certain regular patterns were extracted from the complex in the 1930s, but it was not until after the war, when research into computers boomed, that equipment was devised which could decode the trace as it was recorded, sorting out various regular rhythms from the whole. Three 'basic rhythms' were identified by the end of the war, and a course of experiments to study them was begun immediately afterwards.

It was found that the electrical activity of the brain varies considerably from person to person. The most obvious brain rhythm, in terms of the area of the brain where it can be picked up and the amplitude of the oscillation, is the alpha rhythm, whose pulse-frequency is about 8–13 cycles per second. But the alpha-rhythm behaviour of the brain shows very marked differences in various subjects.

In about 60 per cent of subjects tested, the alpha rhythm manifests itself when they are relaxed, with their eyes closed. When the eyes are opened, or the mind is engaged by asking it to solve a mathematical problem, the rhythm disappears. People whose alpha rhythms behave in this way are classed as R-type (R is for responsive). About half the remaining subject show no alpha-rhythms at all, and are therefore classed as M-types (M for minus). The rest have alpha rhythms which persist even when the eyes are opened or the mind becomes engaged in active thinking. These are P-types (P for persistent).

It is interesting to note that the alpha rhythm seems usually to be correlated not with brain function, but with *lack* of brain function. It has been suggested that the rhythm is associated with the brain scanning the sensory environment for input – rather like a radar antenna going round and round. In order to account for the various types of alpha activity it has been suggested that P-type people tend to think non-visually, and that their actual thought processes do not require sensory support, so that scanning can continue. Corollary to this, of course, is the assumption that M-type people tend to think in visual images all the time. The R-type person (the commonest type) strikes a balance between thinking in visual analogies and pure symbols.

In recent years an interesting series of experiments has been carried out in which subjects are taught to control the alpha rhythms in their brain by 'biofeedback training'. The electroencephalograph allows subjects to *see* what is going on in their brain, and awareness of what is going on inside them makes it much easier for them to control it. Subjects asked to produce alpha rhythms or stop them learn to do so fairly quickly when they can see whether or not they are succeeding. Many subjects who learn to switch their alpha rhythms on and off at will observe that the on-state correlates with a feeling of serenity and detachment − calmness without drowsiness.

The mental state which can be induced by switching on alpha rhythm has been likened to the mental states achieved through the training programmes of 'transcendental meditation', and it has been suggested that biofeedback training offers a short cut to the kind of mood control used by the Zen adepts who practise and teach transcendental meditation. Biofeedback training is now being extended to the other brain rhythms, to see what kinds of self-control can be discovered when they, too, are harnessed to the will.

The second brain rhythm to be discovered was the delta rhythm. When Berger had initially announced his discovery another physiologist, Golla, had been quick to suggest that organic disease would cause disturbances in the brain rhythms, and so it proved. (And thus the electroencephalograph became useful in the diagnosis of certain physical dysfunctions and in the location of brain tumours.) Delta rhythms arise in regions of the brain which are compressed, or invaded by growths, or distorted by injury, and thus presumably reflect the 'switching off' of functions within particular areas. Delta rhythms also appear during deep sleep (i.e. when the subject is not dreaming) and are normal and dominant in babies during the first year or so of life. The frequency of delta rhythms is about 0·5–3·5 cycles per second.

The middle range of frequencies, between delta and alpha (about 4–7 cycles per second) was found to be characteristic of a third kind of rhythm, of smaller amplitude. This was named theta rhythm, and was found to be associated with the emotions of subjects, particularly with pleasure or pain.

As with the alpha rhythm, there are striking differences between theta behaviour in different individuals. In some people theta activity seems to be correlated specifically with pain, in others specifically with pleasure, while in others theta activity is manifest in connection with either. In all instances it is associated with *change* of mood rather than with the mood itself. It can be invoked in various subjects by sticking pins in them, offering a reward, or by taking away a toy. In all cases the mental 'change of state' may call forth a burst of theta activity which may then be damped as the subjects 'control themselves'. Theta control by biofeedback training appears to allow subjects to increase their mental efficiency, enhancing memory and the facility with which they can calculate. This may be because learning to control theta can make a subject less vulnerable to the sudden rushes of frustration or petulance which may impair the efficiency of rational thought – it encourages the subject, in fact, to greater detachment from his emotions.

Beyond alpha rhythms in the 'brain wave spectrum' are beta rhythms, varying between 14 and 28 cycles per second. These are the rarest of the characteristic patterns of brain activity, and it is only occasionally, and for fractions of a second, that the higher part of the frequency range is ever manifest. It has been found that heavy smokers generate a great deal of beta activity and virtually no alpha, whereas non-smokers are often P-type with respect to alpha. It has therefore been suggested that beta is evidence of high tension, and that the elimination of beta through biofeedback training may help to alleviate peaks of mental tension.

It is important to realise that brain rhythms are the effects, *not* the causes, of various states of mind. Attempts to correlate delinquent behaviour in children with theta activity in the brain, which were made after demonstrations of the association between theta and anger or frustration in some subjects, proved negative.

Brain-wave patterns tend to change quite dramatically with age. After the first year, when the dominance of delta rhythms declines, alpha and theta appear, but almost never simultaneously. Theta is generally more common, but a trace with no alpha at all – moderately common in adults – is very rare in children, though on the whole alpha activity is much less noticeable in young subjects. Averaging from a large number of subjects, who differ individually, it seems that theta activity is dominant from one to five, with alpha becoming gradually more dominant thereafter.

The versatility of the brain rhythms, which are correlated with different changes in mental state in different people, is most significant in its implications about the way the structure of the mind develops. The theta rhythm is not in itself anger or frustration or pleasure or pain, but appears to be more like a kind of mental 'clutch' engaged when the mind undergoes an emotional change of gear. The fact that it is manifest in various different contexts suggests that we have to learn to feel such things as anger and pleasure. Physiological reactions to stimuli are inbuilt, but it appears that psychological ones are not – or, at least, that such psychological responses as are instinctive can be overridden and consciously controlled. This fits in very well with the account of human evolution which I offered in the last chapter. The fact that the brain *learns* to use theta may be responsible for the fact that a masochist may find pleasure in pain, and that certain religious ascetics have transcended pleasure and pain altogether. A much more basic implication of this idea is that the way I experience pleasure or anger in my mind is not necessarily the same way that you experience pleasure or anger in yours – and this might help to explain a good deal about different people's different emotional priorities and responses.

The gradual replacement of theta-dominance by alpha-dominance which tends to happen in growing children once they have left the delta-dominant phase where, presumably, they are ruled largely by instinct, suggests that one system of order-perception by which the child relates to the world is replaced by another. The making of choices necessitates sensory information being categorised as it comes in, and structured as it is stored, thus becoming a frame of reference for future decisions. The first system by which the child learns to bring order into the world it perceives is emotional (theta-correlated), but this tends to be replaced, to a greater or lesser degree, by another system – the rational (alpha-correlated) system. It is, of course, commonly observed that young children make decisions emotionally, but as they get older tend to become more rational – it is very rare, though it does occasionally happen, that an adult becomes quite unemotional or remains completely dominated by emotional considerations. The alpha and the theta rhythms, by providing indicators of changes of mental state, may be the means by which the child, at various stages in its upbringing, learns different ways to sort incoming data and regulate its own reactions thereto.

The things which we tend to categorise as 'mature behaviour' and 'childish behaviour' (whether observed in adults or in children) tend to involve alpha-correlated brain activity and theta-correlated activity respectively. An adult's overreaction to pain, quickness to anger or over-indulgence in pleasure is usually labelled 'childish', while the ability of a child to remain calm and relaxed is often taken as a sign of 'maturity'. The brain rhythms may be the basis of the mechanisms of human choice. Perhaps, in the way we learn to use theta activity within our brains we may recognise the mental forces which Freud labelled the id, and in the way we learn to use alpha rhythms we may find the origin of the superego − the ego, of course, being the personality synthesised from the two. If this is true, then biofeedback training may provide an important means of psychological self-readjustment and self-repair.

The electroencephalograph provides us with a means of observing the brain at work − and, to a certain extent, the ability to experiment with it. We can see certain things going on which we can relate to certain attributes of mind, but what we cannot see is the working of the mind itself. Again, we must beware of the intellectual trap into which so many scientists tend to fall − the assumption that what we can see is all that there is.

The relationship between mind and brain is a very difficult one to describe or to understand − there are many attitudes to it deriving from different avenues of approach to the problem.

Descartes, as we have seen, saw the body as a machine which was simply being used by the soul as a temporary convenience. In more recent times this kind of strict dualism (i.e. the complete conceptual separation of mind and brain) tends to be an echo of behaviourist thinking − the machine becomes the essential thing while the mind is relegated to the status of a tenuous ghost unworthy of consideration. For the first dualists, therefore, the mind was the driver of the machine, for later dualists it has become the passenger. In some analogies it may even become a 'back-seat driver', paying close attention to the road ahead, shouting warnings and instructions, but in the final analysis quite impotent to affect the course which is being steered.

All these analogies fail to recognise the unity of mind and body. The essentially religious approach of Descartes is committed to the belief that the soul can exist quite independent of the body, and the behaviourist approach is committed to the attitude that the behaviour of the organism can be studied objectively without recognising the existence of anything else. We must, of course, be careful to distinguish the working of the brain and the working of the mind, because the two are not identical − but on the other hand we must not ignore the fact that the two are fully integrated, that they are, in a sense, different aspects of the same system.

It is the brain that actually does things − receives and processes sensory data, despatches messages to the motor nerves to initiate and control bodily action. The mind is the organisational system which determines *how* data is handled and what to put into the messages which are despatched.

In an earlier chapter, I compared the organisation of the cell to a factory – the cell is the basic production mechanism and over the centuries human methods of production have come to imitate its organisation fairly closely. By a similar analogy we may compare the mind–brain system to the methods of information-handling we have developed in our offices.

Sensory information coming into the brain is like mail coming into an office. It is scanned (the letters are opened and read) and the essential information (the message-content of the letters) is sorted out from the non-essential. Ultimately, all the information will be either stored in the memory (filed) or forgotten (thrown into the wastepaper basket), but in the meantime some input requires immediate reaction while some can be temporarily left to one side to be dealt with later (some letters are dealt with promptly, others are shunted aside for attention in due course). Other information has to be extracted from the memory in order that input can be seen in context (the appropriate file has to be produced for reference in each case). Occasionally, unusual input requires innovative action, and this may lead to confusion (some letters cannot be dealt with by customary procedure, and have to be handled as unique cases).

We all know, of course, that modern offices are hopelessly inefficient, bogged down by a flood of bureaucratic nonsense while the staff care little for what goes on outside. We are also familiar with such observations as are contained within 'Parkinson's Law' – for instance, that work expands to fill the time available for it. Could there be any more conclusive evidence of the fact that offices are reflecting by analogy the ways in which our individual minds tend to work?

The difference between the human mind/brain and, say, the bird mind/brain can perhaps be made clear with the help of this analogy.

Birds, it seems, cannot see beyond 'mere appearances'. Gull chicks in the nest will open their mouths wide for any piece of wood painted the same colour as the adult's beak, and if it is more brightly coloured, may respond even more fervently. Birds learn to 'recognise' their mother soon after hatching – the image of the object that they see most frequently becomes 'imprinted' in their minds and connected to a whole series of instinctive behaviour-patterns. Many birds, however, if they lack a mother, will imprint to almost anything – a coloured balloon or a human being. Some modern buildings presenting vast fronts of sheet glass occasionally have trouble because birds fly into the glass. This trouble is easily averted by pasting paper shapes which approximate to hawk silhouettes on the windows – such a signal is quite sufficient to trigger an avoidance reaction in the birds.

The kind of system which regulates this behaviour is rather like a primitive computer-run office. Incoming letters are scanned for certain key words which evoke certain specific responses. There is little or no reference to files and no capacity at all for innovative action to deal with unusual data. There are many popular anecdotes concerning the impossibility of getting any sense out of computers which operate on behalf of mail order firms and banks. When they send out bills for £0·00 they expect to be paid, and will keep on sending out bills and threatening letters until they are (but if, of course, you finally give in and send them a cheque your bank will complain because their computer is not equipped to handle cheques for £0·00 ...). We all know, of course, that computerised offices are supposed to be much more efficient – they are automatic. But automatism has its drawbacks.

Birds, like computerised offices, are only as good as their programming. They can learn, to some extent – but only one thing at a time, by such processes as imprinting, when new links between input and output are forged. But birds, when scanning their sensory input, can only react to a series of programmed triggers. They cannot do what human office staff are supposed to do and extract *meaning* from the incoming data – recognise patterns within it and degrees of similarity between situations. Nor can they innovate when input turns up which they are not programmed to deal with. When sailors first landed on Mauritius the dodos simply sat around and did nothing. The sailors took to killing and eating them, and still they sat around. Now we have a common simile: 'as dead as the dodo'.

Most mammals are much better at learning things than birds. They make rather more use of their filing systems, and though it is certainly not true that elephants never forget or that Androcles could rely implicitly on the good will of his lion, we have tended to give appropriate acknowledgment to the fact that mammals are not without intellect altogether in the stories we invent about them. Some mammals are also capable of a fair degree of innovation in their behaviour, and even of creative thinking. A chimpanzee may use a stick with which it has been playing as a tool to reach food placed out of reach beyond the bars of its cage, and one experimenter observed a particularly clever chimp actually make a tool by breaking a branch off a bush. These are fairly trivial concrete examples of a generalised ability which many mammals have to vary their behaviour within flexible limits set by their programming.

What the human mind has that the ordinary mammal mind has not is not simply *more* office equipment but a rather more efficient way of using it. Not only are its operating procedures more highly developed, but so are its facilities for stepping outside those procedures in order to deal with particular situations. Such information-handling sophistication as an animal possesses has to come about through the testing (by natural selection) of genotypes, but the human mind can run its own testing procedures, trying out courses of action and recording the results in the files. The most important thing of all, of course, is that human beings can communicate the contents of files to one another: they can use inter-office memoranda, which (despite their nuisance value and their tendency to proliferate far beyond the necessities of communication) are extremely useful.

Animals make very little use of 'inter-office memos' (i.e. language), and when they do they tend to circulate 'ready-printed' messages from a limited range rather than making up their own.

But perhaps the most crucial difference between the office-organisation of animal minds and that of human minds is the lack of a useful and comprehensive system of cross-indexing. It is here that the actual size of the human brain becomes important, for it is the size and the complex cytoarchitecture of the brain which allows elaborate cross-referencing to be made. Most of the mammals which keep and use memory-files can only relate input data to the contents of one particular file — such behavioural adaptability as they have is limited to the method of choosing which file is relevant to the particular input data. The chimpanzee which broke the branch from the bush in order to reach out for its food was, however, going one step further in the use of its filing system — it was relating the contents of one file to the contents of another; it made a cross-reference.

This is perhaps the true power of the human mind — the ability to bring disparate fields of experience together and discover new courses of action by analogy. This is one of the fundamental processes of logical reason — the syllogism, by which two statements may be combined in order that a conclusion may be drawn which contains information not supplied by either of the statements alone.

We all know that if we relax and let our mind drift, a 'train of thought' develops which may carry us from one idea to another, from one memory to another, in an endless sequence whose links may be forged by verbal associations, by visual associations or by temporal associations. There are an almost infinite number of routes through the maze of the memory, and the one which is followed by any particular train of thought may be modified by the context of the initial thought, or even by pure chance where either of two links are equally likely to be made. This kind of process is reflected in word-association party games, and in word-association games played by psychologists who use the kind of ideative connections drawn by a subject's mind to deduce something of the overall mental context (the 'state of mind') in which the associations are being made.

A mind may be said to be as good as its ability to make and remake these connections – especially to create new ones. Many highly intelligent men are fascinated by word-play, and the sense of kinship between the theorists of modern physics and the world of Lewis Carroll's Alice is, in a strange way, a kind of testament to Carroll's ability to make new ideative links with his brilliant word-play. Edward de Bono, champion of 'lateral thinking', relates the power of creative thinking to the ability to make novel cross-references in the mind, and Arthur Koestler, in his analysis of *The Act of Creation*, has argued that the heart of creative thinking is the act of 'bisociation' – the deliberate intersection of different 'planes of thought'. Creative genius is so often undervalued in its own age and environment very largely because most of us have to *learn* to make the kind of connections which spring unbidden into the mind of a Shakespeare or an Einstein. What we can perceive, with the aid of hindsight, as brilliant, often seems at first to be merely strange and alien. The history of art and art-appreciation makes this abundantly clear.

Offices tend to work in a state of barely controlled chaos. So do minds. Within that ever-incipient chaos is the potential for new discoveries, and the real basis of the 'superiority' of the human mind.

Anyone who has ever worked in an office knows that the work never actually gets done. There is always more coming in, and the ideal state of being is merely to have everything under control, so that everything is at least in the process of perpetually being done. This dynamic system has its limitations. The situation may frequently arise where everything actually coming in is being dealt with, so that the everyday business of the office is carrying on, but in the meantime a great deal of the 'back-up work', involving bringing files up to date or re-organising the material within files, or keeping the records up to date, is being neglected because of lack of time. This gradual accumulation of minor organising tasks ultimately begins to threaten the efficiency of the office, and it is periodically necessary to stop dealing with incoming mail, etc., while the whole system is 'tidied up'. Normally, this business of 'getting the office straight' takes place during periods when, for other reasons, input of material tends to be slack.

The mind suffers from this kind of organisational drawback too. Sensory information comes in, is dealt with and acted upon with considerable efficiency, but much of the data is hurriedly stacked in a short-term storage system, without being fully redistributed within, and integrated into the permanent filing system. Periodically, it becomes necessary for the mind to stop dealing with sensory data and get itself sorted out. This is done during the periods of bodily inactivity we call sleep – as we have all observed, the mind is occasionally quite active during sleep in dreaming.

Sleep is very necessary to humans. People who try to go without sleep for any considerable period of time show a very considerable loss in the operating efficiency of their minds, until they get to the point at which sensory data can no longer be handled at all. Some people get by with very little sleep compared to most of us, but they cannot get by with none at all, and they dream more frequently and for longer periods when they do sleep.

The various phases of sleep, as revealed by the encephalograph, demonstrate that the winding down of the body and mind is really quite a complex process.

As we have already noted, when the subject is simply relaxing, with eyes closed, the characteristic brain rhythm shown in the trace is alpha. As he begins to get drowsy, however, the alpha rhythm begins to fade away (thus confirming that it is a scanning pattern, not a symptom of brain inactivity). It is replaced by something which looks similar to (though not identical with) theta rhythm, which begins in the back and sides of the head and spreads gradually over the entire cerebral cortex. At this stage the subject is quite easily aroused by a noise or a touch, and if he is startled he may occasion amusement by promptly denying that he was asleep.

As the body relaxes further, beyond the point where the brain reacts to slight sounds or other gentle stimuli, slow, irregular waves appear on the trace, periodically interrupted by short bursts of smaller, faster 'spindle-waves' with a frequency of about 14 cycles/sec. This level of sleep is the regime in which dreaming occurs, and the bursts of spindle-wave activity correlate with movements of the eyes, as though they were scanning without there being anything actual to see. This is known as REM (for rapid eye-movement) sleep.

In deep sleep, where such interruptions do not seem to occur, the encephalograph is dominated by the slow delta rhythm. In this regime loud sound or firm touches may trigger a complex brain wave discharge called the K-complex, whose effect seems to be to muffle the stimulus and maintain the state of sleep. Steady, repetitive sounds do not interfere with the delta rhythm, but their cessation may well call forth a K-complex reaction.

These sleep patterns describe what happens in the majority of subjects, but individual traces may show as much variation and personal idiosyncrasy as waking patterns. It is interesting to note that hypnosis is not at all like sleep so far as its effects on electroencephalograph traces are concerned. A hypnotised subject told that his eyes are open when they are, in fact, closed may believe it – but his alpha rhythms do not. Hypnosis seems to affect the mind without reference to the brain-activity on which the mind's structure is based. Within the brain of a hypnotised subject stimuli are exaggerated rather than damped or counteracted, as in sleep.

Sleep almost certainly fulfils more than one function. Deep sleep, when the mind appears to be switched off, is probably associated with the restoration of the body to physiological readiness, although why this should not be done while the mind remains active is not clear. While some dreaming almost certainly serves the office-tidying function which I have described this is not necessarily the only purpose of dreams.

If we adopt this concept of dreaming it becomes easy to understand why Freud and many other psychoanalysts have found dreams a particularly fruitful source of information about their subjects. As in the case of word-association, the kinds of connection made in dreams (i.e. the kind of cross-indexing which is constructed while the files are being tidied) can give a strong indication of the kind of mental context in which the work is being carried on. Dreams may be additionally useful because they deal very largely in unconscious connections – manifest in the conscious mind only in symbolic terms – rather than conscious ones.

The explanation of the utter strangeness of many dreams and their apparent lack of logic may have something to do with the symbolic expression of unconscious connections, or it may simply be a kind of 'shuffling mechanism' whose role is analogous to the role of sexual reproduction in genetic reassortment. This kind of random recombination may be an in-built source of human creativity – a number of notable inspirations have come to scientists and artists in their dreams, and commonly accepted good advice to people with problems is to 'sleep on it'.

Christopher Evans has attempted to build a theory of dreaming by analogising the tidying-up process which goes on in the mind to the periodic tidying-up of computer programmes when 'bugs' are extracted and better subroutines introduced. He has suggested that the fantastic aspect of dreams comes about because of interference from new input. He points out, quite rightly, that the dreams we remember are those into which we wake, and which we therefore disturb by new input from the conscious mind. Again, this is creativity deriving from the bringing together of different contexts.

Some new and very interesting data on dreaming have recently been provided by a French psychologist, Jouvet, on the basis of some experiments with cats. Jouvet was interested by the fact that the activity of the brain during dreaming seems not so very different from its normal waking activity, save that all the motor nerves seem to be 'disconnected'. He discovered that the 'switch' which accomplished this disconnection was a small body at the base of the brain called the pons. When he removed the pons surgically from a number of cats he was able to watch them 'acting out' their dreams.

The things which Jouvet's cats did in their sleep were not so very different from the kind of things cats do when they are awake – specifically, the kinds of behaviour-pattern which are instinctively inherited. They stalked and caught imaginary prey and reacted to imaginary predators. It seems, in fact, as if the primary function of dreaming in cats is to allow them to rehearse and practise their instinctive behaviour-patterns. Dreaming may thus be a vital part of the mechanism by which behaviour-patterns are transmitted genetically.

If this is so then the arena of dreams may well be the 'melting pot' in which learned behaviour is integrated into genetically programmed behaviour in the higher vertebrates. Each particular programme might be continually changed (both 'debugged' and 'sophisticated') as the animal gets older. In man, of course, the dream-input from instinctively programmed behaviour would be very much more limited and malleable, and one would expect to find something going on which is much more like the office-tidying process I have described. We now know that the chemical basis of memory is associated with certain kinds of nucleic acid — specifically, with molecules of RNA. These are the same kind of molecules involved in the genetic system of protein-making, and this argues very strongly that memory and the faculties of the mind arose through the evolution of adjuncts to the genetic system of heredity.

We may, perhaps, be able to take this line of argument a little further and suggest that the theatre of dreams provides us with a medium in which alternative patterns of behaviour may be tried out in 'mock-ups' of real situations — gives us, in fact, the ability to make thought experiments in behaviour. By this means alternative reactions might be compared, sorted and selected before a situation is ever encountered in reality. We all know, for instance, the tendency which we have to rehearse anticipated situations consciously — we may 'rewrite the script' for an important interview many times, deciding what to say and how in reply to a whole spectrum of possible questions.

If we do this kind of thing unconsciously, in our dreams, we may be able to find a way to account for the 'disturbing dreams' — comedies of embarrassment rather than actual nightmares — which are among the commonest dreams people have and remember. A recent survey discovered that one of the most frequent dreams recorded by British housewives involved having the Queen turn up unexpectedly for tea when there was no food in the house. Freud observed that one of the most common dreams involved being naked in public. The disturbance which such dreams bring about may be the way that behaviour-patterns are 'failed' by internal experiment, so that in everyday life we always take the trouble to get dressed when we get up and housewives always get the shopping in, no matter how tedious and pointless these acts may occasionally seem.

It is perhaps in this way that dreams may be useful in revealing worries and obsessions, if particular behavioural experiments have to be carried out over and over again while the lesson still does not stick. Certainly, it is reasonable to suppose that the plays which are acted out in the theatre of dreams may be an important mechanism in the 'social conditioning' which we all receive covertly, and which teaches us to know which things are 'not done' in spite of there being no obvious or rational reason. Perhaps the work of Jouvet and the kind of theory which Christopher Evans has built will allow us to put Freud's massive work on *The Interpretation of Dreams* into its proper context at last, so that the ideas therein can be overhauled – and its programme for psychoanalytic action debugged.

At present, our knowledge of the mind remains confined by a cloak of ignorance. The theories by which we seek to describe its properties and account for its manifestations are extremely primitive. The mythology of the mind is the most intuitive and superstitious of all the mythologies of modern science.

In the physical sciences, lack of knowledge is generally paralleled by lack of practical power. The theory of the steam engine and the steam engine itself went hand in hand. Atomic theory and the atomic bomb grew up together. In the science of the mind, however, this is not so. Though we do not understand the mind fully, we nevertheless have means of altering it and meddling with it which are quite considerable.

The power of matter over mind has been recognised since prehistory – as has its hypothetical counterpart, the power of mind over matter. Intoxication was, at one time, regarded as a transcendental experience – a gateway into a divine world. Dionysus was the god of wine and ecstasy in the Greek pantheon. In the Roman Catholic mass, wine becomes the blood of Christ via the miracle of transubstantiation. Cannabis, *Amanita muscaria*, peyote and many other plants with psychotropic (mind-changing) effects have been used by the people inhabiting the areas where they grow since time immemorial, often in connection with religious belief and celebration. In Europe belladonna (deadly nightshade) and bay leaves were added to alcohol in order to promote the transcendental experiences associated with witchcraft.

Thus, we have always had the power to interfere with our own minds – and, to a limited extent, with the minds of others – in a more or less strategic fashion. Science has not changed this basic ability – we do not know enough about the chemistry of psychotropism to increase the strategic element of the interference very much, if at all. Science has given us purer drugs, and a wider range of drugs, but it has not given us the capability to organise the power of matter over mind, or even to understand it. And yet there are a whole host of new problems connected with drug use which simply did not exist a hundred years ago.

Basically, these problems arise not because of new discoveries in science, but because of new contexts for thought and action provided by science, and in particular by the growth of scientific medicine.

Science has brought about a revolution in what we think of as 'health'. In what we think of as 'primitive societies' the whole business of illness and healing is conducted in a manner very different from our own. Illness is countered not only with drugs but also with ceremonies – it is seen not as a physical disorder caused by some agent extraneous to the self, but more as a failing *of* the self, belonging not merely to a scientific context but to a religious and a social one. Cure, in such a context, is not merely a matter of treatment, but of faith, of self-confidence and of the repair of social relationships.

In the Western world, science fought a long, hard battle to translocate the human being from a religious context to a scientific one, and in the nineteenth century it succeeded. It is no coincidence that the triumph of Darwinian philosophy was accompanied by Pasteur's revolution in medical science. The fight to replace medical superstition by medical science was long and bitter, and cannot honestly be said to have been won. The promotion of medical miracles is still big business. Laxative panaceas, brand-named aspirins and placebos continue to be effective in restoring the self-confidence of the public with respect to a great many of their troubles.

Medical science has its heroes (Pasteur, Koch, Lister, etc.) and also its martyrs (Ignatz Semmelweis and Florence Nightingale), and the story of its fight against the miseries of human disease can be made into rousing reading. Nevertheless, 'quacks' and 'charlatans' continue to flourish, although they have been forced, to a large extent, to adopt the trappings of science: machines, jargon and mock-logic. This fact has forced scientists to admit that a certain amount of illness – no one knows just how much – is perfectly amenable to cure by just about any process which a patient believes in. They are careful to distinguish between organic diseases (which have real, physical causes – usually bacteria or viruses) and psychosomatic ones (whose physical symptoms are real, but which have imaginary causes).

The fact that medical science has been forced to make this distinction is really tacit recognition of the fact that the 'primitive societies' are quite right, and that illness does not belong exclusively to a scientific context – that not all cures are treatments, and not all treatments cures. Primitive societies may have very limited scientific resources for the treatment of organic diseases, but ignorance is not necessarily correlated with stupidity.

It has been recognised, therefore, that psychosomatic illness does not belong to the same scientific context as organic illness. In consequence of this, rationalist philosophy has attempted to provide an alternative scientific context to which it does belong.

We have, in the twentieth century, developed the concepts of 'mental health', 'mental hygiene' and 'sanity'. All of these concepts have been developed by analogy with organic medicine, and are wholly artificial, no matter how useful they might be. It is this branch of medical science to which the class of events connected with psychotropic compounds belongs. The ability to tamper with the mind has been translocated from its religious context (where 'madness' itself once belonged) to a new context of *mens sana in corpore sano*. Drug users and opponents of drug use alike tend to accept this context, and most of the arguments about the problems of modern drug use merely involve the roles attributable to drugs within the context, not whether they ought to belong to it at all.

The ability to tamper with the mind is no longer confined to chemical means. In 1949 W. R. Hess conducted a Nobel prize-winning sequence of experiments demonstrating that many functions of the organism may be influenced by electrical stimulation of the brain, including fear and anger. Other similar discoveries followed quickly, the one which captured the public imagination being the discovery by Olds of the misnamed 'pleasure centre'. Rats given the means to stimulate themselves in this particular area did so continuously, ignoring hunger and thirst to a large extent, and stopped only when they dropped from exhaustion.

At about the same time the discovery of ataractic drugs (generally known as 'tranquillisers') increased the range of effects which could be induced in the mind chemically.

Ataractics quickly became commonplace. They were taken up by the medical profession in a big way as the first important scientific servant in the treatment of psychosomatic illness and general mental/emotional malaise. The extent to which the use of drugs is now taken for granted, in much the same way that the use of aspirin as an analgesic is taken for granted, is eloquent testimony to the fact that we recognise no difference in context between the different ranges of phenomena with which these drugs are supposed to deal.

Since the launching of the ataractics there has been a veritable flood of psychotropic drugs invented and recruited for medical use: sleeping-pills, pep-pills, anti-depressants and pain-killers. Other drugs often have psychotropic side-effects – for instance, antihistamines used to relieve allergic reactions often cause drowsiness.

Curiously enough, while this explosion in psychotropic practices has taken place under the aegis of medical science, serious worries have been generated about the psychotropic effects of a range of drugs which, because they are of greater antiquity, have been in common use in society for some time: tobacco, alcohol, cannabis, heroin etc. I think it is true to say that the acceptance of one class of drugs and the bitter opposition to the other is based far more on the context to which they are seen to belong than to the actual effects which they may have. (Addiction to the medically controlled drugs is probably as widespread as addiction to those which are not.) Perhaps the most interesting and indicative case is that of the drug LSD–25, the first of the so-called 'psychedelic' drugs to be developed in the laboratory, which was first discovered and used within the scientific context but then 'escaped' and became a cause for extreme concern.

Many of the laboratory-developed drugs were initially regarded as being connected with the chemistry of mental illness, and in particular with 'schizophrenia'. This is one of the principal categories of mental malaise, and its 'symptoms' have a degree of consistency, although to imply that we can talk about schizophrenia as specifically as we can talk about cholera or malaria would be highly misleading. LSD was originally considered a 'schizophrenia-inducing' drug, and the ataractics were initially promoted as representing a great breakthrough in the treatment of schizophrenia. We can see here the evidence of great faith and optimism on the part of medical scientists who believed that a science of mental health could be developed which would be exactly analogous to the science of physical medicine.

In the name of that optimistic faith, patients in mental hospitals have over the last couple of decades been subjected to a barrage of new treatments, involving the host of psychotropic drugs and the techniques of electrical stimulation developed in the wake of Hess's work. The degree of success which has been achieved is, to say the least, unimpressive. All the treatments have enjoyed some victories (mostly temporary ones) − but, as has already been pointed out, *any* treatment will enjoy a degree of success simply through the faith of the participants in the experiment. No miraculous changes analogous to the transformation in the lives of diabetics brought about by the discovery of insulin have ever been recorded in the medicine of the mind.

Tests which are carried out on new drugs concentrate almost exclusively on organic side-effects. The question asked is generally: does it cause cancer? (Cancer being the bugbear of modern organic medicine.) The effects which the drugs have on the mind is, for the most part, taken for granted − as Herbart observed, how can you conduct proper experiments with the mind? One reads in a number of popular accounts of the magnificence of medical science that the day will very soon dawn when we can buy our state of mind at the chemist's shop; happiness, intelligence, virtue and fear will be available on prescription. Perhaps this is an exaggeration born of optimism, but we have already taken the important conceptual step − today, we regard unhappiness and misery (jargonised as 'depression' and 'melancholia') as diseases just like the common cold and bronchitis. People who suffer from them go to their doctor. He might not be able to give them a prescription for happiness or joy, but they have been led to expect results and that is what he tries to deliver. He offers them tranquillity or zest in the shape of ataractics or amphetamines, because those are what are immediately available.

What the long-term effects of these drugs may be we do not know − we have had no long term to study them. We are not even in a position to say with real confidence what their immediate effects are. The science of mental health does not, in any practical sense, exist. It exists only because we believe in it −

because we are quite convinced, as the inheritors of nineteenth-century rationalism, that there must *be* such a science, and that we will find it all the faster by pretending we have it already. In the meantime, our assumption that 'mental health' and 'physical health' are directly analogous, and that the procedures appropriate to the handling of organic disease are exactly similar to those appropriate to handling psychosomatic disease, is both unfounded and highly dangerous. But it is, perhaps, understandable.

When Haeckel set out to answer Bois-Reymond's seven world-enigmas in *The Riddle of the Universe* he was not really interested in the questions themselves. What he wanted to do – and what he felt compelled to do – was to insist in no uncertain terms that to this set of questions, and to any other set that any other man might care to pose, there *were* answers, and that those answers were – and had to be – provided by SCIENCE. His book was an expression of profound faith. The actual answers which he put forward seem to us today to be highly unsatisfactory – he got his sums wrong because he thought he knew so much and really knew so little. In the twentieth century we have discovered the errors in Haeckel's thinking, but we have not really given up the assumption behind it.

We now know that there are limits to the answers which science can provide. We now know that there are limits to the things we can know – what we can perceive is only a very thin slice of what *is*. And yet, to a large extent, we persist in assuming that the answers to our questions are there to be found, and that SCIENCE will do the job, tomorrow or the next day. There are uncertainties about the subatomic world, mysteries concerning the far reaches of space, but perhaps this does not matter so very much – these are things which do not directly concern us. But the human mind does concern us – it is, in fact, our most immediate concern. Here, as nowhere else, it is important to realise the limitations of science – and, more important still, the fallibility of our blind faith in scientific answers. Faith in science is, itself, quite unscientific. We must not assume that theories applicable to one set of entities and events are applicable to another, simply because we need a theory of some sort.

It is sadly true that the biological sciences – and, in particular, the human sciences – have advanced these last hundred years in an intellectual climate of faith rather than inquiry. This does not detract from many real achievements, but it does mean that we must be very careful indeed to distinguish between what we know and what we believe — and think very carefully about the actions and opinions which spring from the latter rather than the former.

The third of the major hypotheses which we have considered in the light of twentieth-century science, the mind (the others, of course, being the atom and the universe), is perhaps the least approachable of the three. It is the least amenable to objective study, the least governable by rules and fundamental principles. In this area more than any other the redundancy of simple answers — whether the 'scientific' simple answers of Pavlov and Watson or the intuitive simple answers of Gall and the astrologers, who attempted to find the map of personality among the maps of ordinary objective phenomena — is obvious and unquestionable.

In the biological and psychological sciences the mysteries of modern science are by no means so esoteric as the mathematical mysteries of the physical sciences: They can be perceived by all of us. Perhaps, though, the most important problems which these sciences face are not a matter of finding solutions to these mysteries, but of recognising the complexity — and perhaps the inaccessibility — of such solutions relative to the current state of our knowledge.

Part Three
Science and Man

Yea, my heart had great experience of wisdom and knowledge. And I gave my heart to know wisdom and to know madness and folly: I perceived that this also is vexation of spirit. For in much wisdom is much grief: and he that increaseth knowledge increaseth sorrow.

Ecclesiastes 1, 16–18

10 The Marriage of Man and Machine

Fortune disposes our affairs better than we ourselves could have desired: look yonder, friend Sancho Panza, where thou mayest discover somewhat more than thirty monstrous giants, whom I intend to encounter and slay; and with their spoils we will begin to enrich ourselves: for it is lawful war, and doing God good service to remove so wicked a generation from off the face of the earth.

So said Don Quixote on sighting the windmills, and despite Sancho Panza's protests he proceeded to attack. He had, of course, no chance. The first windmill unhorsed him.

But was Don Quixote right to attack the monstrous giants? The windmill was one of the earliest examples of automatic machinery, harnessing the power of the wind in the grinding of flour. All tools are the extensions of man — transforming his strength and his gripping ability into specific actions tailored to specific tasks — but the windmill was in the vanguard of a new generation of tools, converting the actions of nature to human purpose. The windmill was truly a giant among machines. (And, for what it is worth, they had been introduced into Cervantes' homeland only a decade or two before he wrote Don Quixote — to him they represented a new technology.)

In later ages there have been other men who have believed with Quixote that it would be 'doing God good service to remove so wicked a generation from off the face of the earth'. The majority, however, have discovered that the way to enrich themselves with the giants' spoils is not to destroy them but to enslave them — and that is what history has seen us do.

We may now look back from the vantage point of today and ask how we have used these giants — and how, from their situation of servitude, they have used us.

There are, basically, two ways that a conquest may be reversed: the conquered may revolt against their masters and hurl them back whence they came, or they may simply absorb the conquerors into their own race. The latter case has been the story of mankind's post-conquest relationship with the machines. The machines have seduced us. There are two kinds of windmill – the one a giant, the other a brightly coloured children's toy.

Today, we are married to our machines. In the developed countries, we can no longer live without them, and we are working hard and earnestly to ensure that the underdeveloped countries, too, become wholly dependent upon them.

With the aid of hindsight, we can now trace the steps which led us into this relationship, and identify the crucial moments in the courtship. If the flirtation began with Quixote's windmills, then the engagement was sealed in the early part of the eighteenth century, when Abraham Darby found a new fuel for use in blast furnaces (coke) and paved the way for a new Iron Age, transforming Europe into a land flowing with metal and money. In the same decade Newcomen pioneered the first practical application of steam power, so that a new generation of machines appeared alongside the metal to make them and the fuel to power them.

The importance of steam power was that it was wholly artificial. The windmill, the water-wheel and the sailing-ship were important in freeing man from the limitations of his own strength, but though man used the wind he could not command it. He remained (and still remains) at the mercy of three of the ancient elements: earth, water and air. But the fourth – fire – he *could* command. Steam engines gave him the means, for the first time, to make fire work in a hundred different ways. With the advent of the steam engine man evolved from the use of natural forces to the use of artificial ones.

In 1764 Hargreaves' spinning jenny began the technological revolution in the textile industry, and was quickly supplemented by Arkwright's water frame and Crompton's mule. Cartwright's power loom extended the revolution into weaving in the 1780s, and wool and linen joined cotton as the produce of an Industrial Revolution.

In 1785 a button-maker named Boulton became the first manufacturer of custom-built steam engines. His motto, it is said, was 'I sell what the world wants: POWER!' It is history rather than his own emphasis which has transposed the final word into capital letters.

By this time the electric current had already been discovered, and in 1795 Volta invented the electric battery, thus taking the first important step in harnessing a new power; not a new *source* of power but a new medium by which power could be put to work.

The final year of the eighteenth century, 1800, saw the emergence of mass-production methods in Whitney's musket factory. The importance of this step can best be judged if we make use of the analogy we have already drawn between the factory and the cell — especially the egg cell. Making machines bigger and better and more able is really a secondary priority in evolutionary terms; it is a means to the end of making more and faster, in the same way that making organisms is only a means to the end of making eggs more efficiently. What Whitney did was to standardise the parts of his muskets so that a whole, working machine could be assembled from any complete set, so that each part no longer had to be tailored to its specific counterparts. This was the origin of mechanical 'clones'.

In 1814 George Stephenson invented the steam locomotive, and gave machines the power of movement.

It was in 1819 that Charles Babbage first conceived the idea of a machine to do mental work rather than physical work. He worked out the principles of what he called an 'analytical engine' and which we call a computer. Unfortunately, the precision of nineteenth-century engineering was inadequate to the task of designing parts to the exact specifications which he drew up. Babbage failed to build a working analytical engine, but in his efforts to do so he helped to found and give a direction to the machine-tool industry: the development of better machine-making machines.

In 1854 Bessemer invented the converter which turned iron into steel and gave machines new, more durable and more powerful bodies. Eleven years later this discovery was supplemented by the discovery of a whole new class of materials, offering tremendous potential for the building of machine-bodies, when Parkes developed the first artificial plastic from nitrocellulose.

The electrical revolution got under way shortly after this, when the principle of the electric motor was discovered, greatly increasing the applicability of the electric current. The integration of electrical devices into mechanical engines was, in a sense, the first step in machine symbiosis.

In 1868 Farcot invented an automatic control for a ship's

rudder which co-ordinated its own actions according to necessity. He named the device a servo-mechanism, and it represented the first step in machine self-sufficiency.

In 1875 – exactly a century ago, as I write – the director of the American Patent Office resigned, so rumour has it, because there was nothing left to invent. He believed, apparently, that the Industrial Revolution had (so to speak) run out of steam. He was wrong. Though the possibilities of the steam engine had been thoroughly explored, the possibilities of electrical power were just about to bear fruit in great abundance. Many of the early electrical inventions were 'ahead of their time' in that they could not be made immediately practical, but their day was soon to come.

Reis had demonstrated the telephone in 1861, and in 1876 Bell mastered the practical application of the device and patented it (winning the race to do so by just two hours). Edison invented the phonograph in 1877, though it was not until the early years of this century that a serviceable gramophone could be put on the market. The real potential of electricity as the motive force for the next generation in the evolution of machines, however, only became clear in the glare of the electric incandescent lamp, which Edison invented in 1881.

We have already seen the significance of Hertz's demonstration, in 1887, of the existence of electromagnetic waves, which led Marconi to the development of wireless telegraphy. We have also discussed the importance of Röntgen's discovery – in 1895 – of X-rays. These were the foundation stones of a new science, but in terms of the evolution of machines 1895 was an important year for other reasons. The first cinema films were made and shown, creating a new medium of entertainment, heralding a new era in human communication. In the same year Gillette invented the disposable razor blade, pioneering a new philosophy of transience and built-in obsolescence. It was correctly predicted that if he could invent something people would use and then throw away he would make his fortune, and many a twentieth-century fortune has been built upon the same principle.

By 1900 – again, the final year of a century – Marconi was ready to produce radio transmitters and receivers commercially. In so doing he developed the means of machine-to-machine communication, using a medium quite imperceptible to the human senses.

And so the twentieth century began, with all the potential for a Machine Age having been carefully developed. Machines, by this time, had duplicated virtually all the properties enjoyed by living organisms: movement, self-government, reproduction, communication. Like viruses, machines were not independent of living systems – they could only function in association with men – but while the association continued machines could manifest all the phenomena characteristic of living things.

In the early years of the twentieth century, machines, in collaboration with the Wright brothers, learned to fly. The invention of the triode valve increased the applicability of electrical devices by permitting amplification of the current and feedback control. The embryonic development of complex machine-bodies was greatly facilitated by the development of the assembly-line system in Ford's factories.

In the 1920s machine communication took entertainment and information services into the home for the first time. While the first radio broadcasts were going out the development of the TV was already beginning. (Curiously, many 'prophets' saw the TV as a logical extension of the telephone rather than the radio set, and its potential impact on the pattern of life was considerably underestimated by all but a few commentators.)

The Second World War saw two crucial developments in machine evolution. On the one hand, a new source of power was discovered – not merely a new tool for the applicability of fire-power, like steam or electricity, but a new means of generating power: atomic fission. Second, the first electronic computer, ENIAC, was built and used, ushering in a new era in machine sophistication.

It was the development of atomic fission which demonstrated once and for all the power of the technological giants. The little tailor killed seven flies with one blow and then went out into the world to terrify giants with his prowess. The atom bomb dropped on Hiroshima killed 100,000 people in one blow, and little tailors the world over knew that they had met their match in a new generation of giants.

The impetus given to technological research and development by the war continued in the 1950s. The atom bomb gave way to the hydrogen bomb. ENIAC, with its thousands of clumsy valves, was rapidly rendered completely obsolete by the invention of the silicon transistor, the development of cryotronic switches (low-temperature superconductors which increased the speed of information transfer in machine nervous-systems considerably) and the gradual progress in the micro-miniaturisation of circuitry.

Today, machines have evolved not only to duplicate most of the properties of life, but also most of the abilities of man. They remain, like viruses, completely dependent upon their relationship with men, but that dependency should not conceal their considerable 'genetic' complexity and the scope of their programming.

Like cultural evolution, machine evolution is not guided by the process of natural selection. It has, in fact, consisted of a process of strategic mutation. We, the machine-builders, machine-designers and machine-users, have been both mutagenic agents and hosts to the machine. We have developed the machines as artificial symbiotes. As in the case of any symbiotic relationship, either man or the machine might be represented as a parasite, depending upon one's point of view.

However, for better or worse, for richer or poorer, in sickness and in health, the marriage has been made. And, just as the focus and context of interhuman marriage is the home, so the home is also the focus and the context of man's marriage to the machine. 'A house', said the twentieth-century architect Le Corbusier, 'is a machine for living in.' And so it is. Every time a modern man or woman pushes a plug into a socket he or she is consummating the marriage of man and machine.

Ever since man's new partner in life evolved brains – a mere quarter of a century ago – the idea of man/machine rivalry has found a new context. The modern mythology of science fiction is replete with cautionary tales concerning the ability and readiness of machine-minds to replace human minds in commanding power (both mechanical and political). Already we begin to feel that there are 'Machinavellian' forces at work in the world. Computers write letters to us and handle the financial affairs of large companies. Instances of their stupidity have become the basis of a whole new class of bigot-jokes.

Computers are already more able than men in a great many respects. All their abilities are dependent on their programming – but then, so are ours. The difference is that we are only partly programmed by other men (education in the broadest possible sense) and by innate characteristics – the rest (how large a residue remains is open to question) we handle ourselves. How long this difference will continue to be real or important is not known. What computers actually do is controlled entirely by input – but then, so is what we do. *We* have the faculty of choice – but again, how long this difference will continue to be real or important is not known.

Approximately a hundred years ago Samuel Butler wrote his Utopian satire, *Erewhon*, in which the Erewhonians have given up machines because the machines were evolving so rapidly as to be certain eventual winners in the 'struggle for existence' if they were not exterminated while the chance was still there. (The sole Erewhonian apologist for the machines argued that the genocide was unjust because man, also, is a machine.)

Machines do not evolve by natural selection and do not 'compete' in the Darwinian sense, but there are undoubtedly other ways in which machines may 'take over' some of the functions of the human mind and body. Already it is commonplace that we keep time by clocks and wristwatches rather than by our own bodily experience of elapsed time. This is one physiological function which we have already surrendered to the machine. We are also beginning to surrender certain mental faculties – as witness the current boom in pocket calculators. As to physical faculties, a machine is writing this page for me in standardised characters, so that the printer will not be inconvenienced in setting up the type of his machine, which will print the page for you to read. I am not even using my own mechanical power to depress the typewriter keys – electricity does it for me.

The development of the medical cyborg – the man–machine hybrid – demonstrates the capacity which the machine has for undertaking the functions of the body when it is deemed necessary or desirable. It is now commonplace for heart/lung machines to relieve the body of the burden of carrying out its most vital functions during major surgery. Artificial kidneys – dialysers – have been developed, and there are many people whose life depends on the work of these external mechanical organs. In 1962 the 'Scribner shunt' was invented for use with dialysis apparatus – a permanent Teflon/rubber socket implanted permanently in the patient's body which allows him to literally plug himself in to the machine whenever necessary. Many thousands of people have had supplementary mechanical organs added to their hearts – pacemakers – in order to keep those overstressed organs beating. Modern pacemakers are fully implanted inside the body, and various research teams are working to develop radio-activated stimulators to control the working of the bladder and the pressure of the blood in its vessels. Atomic pacemakers which do not need replacement as their batteries expire are under development. Artificial heart valves have been used successfully and whole artificial hearts have been tested (though not, so far, in humans). Stainless steel bones, fully equipped with artificial joints, are proving useful, and entire prosthetic limbs increase in sophistication and usefulness every year.

The man–machine hybrid is already with us, and is socially acceptable in the name of medical science. The implications of cyborganisation, however, obviously go far beyond medicine. If it is possible to give a man who has lost an arm a new one which is stronger, more adaptable and more useful, can it be so very long before people begin to think of having their own healthy arms replaced? The time is upon us when the tools we use may very well become, in the literal sense, *extensions* of man. When we can already equip ourselves with sockets to plug in kidney machines, must we not anticipate a time when it will be possible for us to equip ourselves with sockets to plug in sewing machines, tractors or telephone exchanges in order to control them directly through our own nervous systems?

Butler's machine apologist argued that the human body is only a machine. It is already possible today, albeit in a very small way, for man to participate in the processes of machine evolution as well as the processes of biological and cultural evolution, and in that sense we are – or can be, if we choose – machines as well as animals and men. And the mechanisation of the body is only a step towards the mechanisation of the mind. How much more powerful, intelligent and accomplished might a brain become if it can, through organic/electronic synapses, be wired up in series with a giant computer? The beginnings of mind–machine symbiosis are already with us, in the shape of the techniques of biofeedback training described in the last chapter. Already, it is possible for the mind to assume control over 'automatic' bodily functions with the aid of machines which transmit to it the data which cannot be obtained through the ordinary nervous/sensory network. The mental control which takes years to master through the difficult instruments of religious mysticism may become easily and quickly available thanks to the electroencephalograph. This link between the machine and the human spirit may provide the component in the man–machine marriage analogous to love.

The cyborgs are the children of the man—machine marriage who testify most eloquently to their mixed parentage. But to a lesser extent we are all children of that marriage. It may not be evident in our genes, but as we have already seen, man is much less the product of his genes than other creatures – we are shaped to a far greater extent by our environment, especially our home environment. In that environment – and most particularly in the home environment – machines have a very significant role to play in shaping our expectation with regard to the world, and the kind of strategies we evolve within ourselves to cope with it. We bind ourselves to keep machine-time. We commit ourselves to machine-transport, machine-production and machine-war. We still may feel that our 'true' parents are human and only that, but that machines have been to some extent our foster parents none of us can deny.

Who are these modern children of the machine? In what way are they different beings from the children of men in other ages? If, while man invents tools, tools invent man, what kind of man is man mark XXI – the man of tomorrow, the child of today?

The early prophets of the Messianic machine – Francis Bacon in his *New Atlantis*, L. S. Mercier in his *Memoirs of the Year 2500*, etc. – saw the machine as a means to liberate mankind from labour, and to create a truly independent and free being. They saw the machine as a supplier of physical needs, which would allow men to devote themselves entirely to the life and work of the human spirit. Their prescription for Utopia was, in truth, a prescription (in today's commonplace use of the word) – to supply the needs of the body as a basis for all-round health. This supreme optimism, despite opposition from romantic writers and philosophers, remained viable until the beginning of the twentieth century. J. B. S. Haldane, in his classic essay of the 1920s, *Daedalus*, was still looking forward to the scientific liberation of mankind, though his contemporary H. G. Wells – last of the English Utopians – gradually gave way to despair in becoming convinced that the means to supply all physical needs was by no means enough. In a reply to *Daedalus* called, inevitably, *Icarus*, Bertrand Russell pointed out that though we may have the means of liberation, the use we make of those means is quite opposite to the use which Haldane and his intellectual predecessors would have us make of them.

In the nineteenth century Karl Marx introduced the concept of the alienation of man from the products of his machine-assisted labour. He attributed this alienation to the fact that the worker did not own the machine, and thus did not control the means of production, so that his labour and his life became, like the product itself, a commodity controlled by capital. In the twentieth century, however, we tend to look elsewhere for the roots of modern man's alienation. As early as 1889 William Morris, in his prescription for a socialist Utopia, suggested that it was not merely the ownership of the machines, but the machines themselves, which alienated men from the products with which they filled their environment. Like the Erewhonians, Morris's Utopians abandoned machines and recovered the arts of craftsmanship.

In the twentieth century, the dream of liberation from labour has taken on something of the aspect of a nightmare. If the machines are to take over the task of providing for man's welfare, does not man become rather superfluous? If he is no longer necessary — or even adequate — to the task of supplying his own needs, what is the purpose of life? In today's world unemployment has become a social enemy and the use of leisure time a social problem — not at all the situation which Bacon and Mercier envisaged.

The mechanical 'liberation' of man — even in its present imperfect state — is already creating a new kind of 'free man', but the social adaptation of men to this freedom is not something achieved by a momentary change of mind. We are only just beginning to see the emergence of social adaptations to a culture based on machine-intensive labour. This seems to be taking place largely as the redefinition and recovery of work: the new prominence of 'social work', which is occupying more and more people as the manual labour force declines and the bureaucratic labour force is displaced by computers. The foster children of the machine are becoming more interested in one another and more involved with one another. This is perhaps not easy to see when old communities are being destroyed by the machine-shaped present, and we are undoubtedly in a transitional phase, but new community-structures *are* being created and adapted to the circumstances of the man–machine family.

Perhaps the most important aspect of the re-making of man alongside his machines and the evolution of new kinds of community is the revolution in communications which has enveloped us during this century. This is really a second-stage revolution, the first stage having been the invention of print.

In an earlier chapter I stressed the important role played by language in human evolution. I commented also on the importance of writing as a means of storing and preserving speech – making a paper currency out of the wealth of ideas. Print brought the gift of the machine – prolific production – to the medium of written language (and thus, indirectly, to the 'production' of knowledge). Print allowed the evolution of a whole new literature which – free from the need for oral communication and preservation – became more complex, more detailed, more analytical. In the beginning, print was a medium used only by a small fraction of society, but as the industrial revolution reached its heights in the nineteenth century so literacy spread throughout society in the developed countries. With the advent of universal literacy print became an all-inclusive medium of intracultural communication; a medium for the production and distribution of myth and custom.

Writing and print are media of a particular kind: they involve a high degree of *coding*. In order to extract the information transmitted by writing or print the eye scans a single long line of characters (broken up into smaller sections for convenience) and the mind translates groups of characters into ideas. (I have already drawn an analogy between this process and the process by which a gene is 'read' and molecular groups are 'translated' into amino acids which are formed in a linear array to make protein chains.) The eye – the actual sensory apparatus involved in reading – operates only as an input device with respect to print. There is an important difference between this use of the sense of sight and the way it is normally used in scanning visual images for information – learning about the world by direct observation.

There is the same difference between the use of the ear as an input device for speech and as a direct 'observer' of the auditory environment, but in the case of sight the difference is more extreme. We tend to use our hearing almost exclusively in connection with speech, the most notable exception being the use of auditory danger signals or signals designed to attract the attention – bells and buzzers. This is not true of sight – we make far more use of sight in perceiving the world, describing it and classifying it. Virtually all our decisions and our judgments are made on the basis of seeing. Seeing is believing, we say, while we complain of 'hearing things'. It is that which is out of sight which is also out of mind, and 'I see' can be synonymous with 'I understand'. These phrases provide a good indication of our sensory priorities. (In many animals, of course, the situation is entirely different – the cardinal example being bats, which are nearly blind but which can measure their environment accurately by echo-location.)

When we read, therefore, we are exploiting a speciality of the sense of sight – a speciality which subverts the normal role of sight in coping with the sensory environment. When we read, we 'switch off' our powers of observation – a total change of attitude which is not required when we use our hearing to listen to speech. Reading requires that we ignore our sensory environment, while listening does not, because we still remain sensitive to the bells and buzzers which seek to engage our attention.

This point is crucial to the understanding of why print had the potential to change mankind so drastically, and why the second stage of the communication revolution has the potential to bring such profound new change. Mechanical production may have separated man from the products of his labour in a metaphorical sense, but print, in a very real sense, separated man from his environment. Reading involves *abstraction* of the self from the sensory environment; a withdrawal into an alternative world – a world in which all information is coded in language, and there is no direct perception. There is, all around us, a world which we literally *know by sight* – the perceptions of touch, taste, smell and hearing are only supplementary so far as we are concerned. But there is also another world, available to us since the advent of universal literacy – a world which we enter consciously by giving up 'seeing' in favour of 'reading'. The two worlds cannot be present simultaneously, in the same sense that the dual functions of hearing can be simultaneously served when we are listening to speech.

This divorce between the 'real world' and the world of the printed word is reflected in many attitudes which prevail today. We take it for granted that a 'bookworm' is necessarily out of touch with reality, that a great scholar has necessarily exempted himself to a greater or lesser extent from the business of mundane living. We recognise a disparity between 'book-learning' and practical experience. A fondness for popular literature is often labelled 'escapism' – and, indeed, the way that popular fiction has evolved genres which feature stylised and formularistic fantasies akin to, but strategically different from, the real world is a facility granted by the necessary exclusion of sensory reality practised by the reader.

The changes wrought in man by the advent of mechanical production and the changes wrought in him by the emergence of the print medium were, in a sense, complementary. The steam engine created leisure, the book offered potential for filling it. Ironically, however, no sooner had the print revolution become established – in the late years of the nineteenth century – when a new revolution began, and the divorce of the book-bound world of knowledge and ideas from the sensory world of experience was rendered all but meaningless.

The second stage of the communications revolution was the mechanisation of speech first via the telephone, and later by radio. Unlike the transformation of writing into print this was not simply a matter of facilitating storage, preservation and distribution, though these facilities were developed in parallel in the form of the gramophone and the tape recorder. The essence of speech mechanisation was to render long-distance communication *instantaneous* and to allow a vast number of people distributed over a large area to listen to a single source of information simultaneously. As in the case of print, words were translated into physical form and carried mechanically, but they were nevertheless received as speech, without being coded into characters.

There was no question of the new channels of auditory communication replacing print (or writing) but they added a new facet to information handling and distribution in developed countries. Alongside mechanised speech, however, came mechanised sight; in 1895 the Lumière brothers developed the kinematograph and pioneered the projection of moving pictures on to a screen. Unlike the auditory media – from which, until 1930, it remained separate – film overlapped the functions of print. Like print, the essential faculty of film was the storage and preservation of information, but film stored actual sensory experience instead of coded data. Film preserved exactly what print excluded: the substance of observation.

The great impact of cinema was in the realm of popular entertainment. Like the book, the cinema provided the potential for creating alternative worlds. It created, however, not abstract idea-sequences to be interpreted by the eye, but sensory experiences to be observed just as the real world was observed. There was a great demand for alternative worlds of fantasy – the artificial environments of genre fiction – and this demand the film-makers set out to supply.

Like reading, film-watching demands a divorce from the

sensory observation of the real world, but unlike reading it does not demand that the actual process of seeing be altered – merely that one should enter a special artificial environment: a cinema, a 'picture palace'. Reading involves active concentration, while watching film is essentially passive – the film creates its own sensory world of make-believe. (As we have already noted, seeing *is* believing, while reading requires the deliberate 'suspension of disbelief'.)

Film did, to some extent, replace print. It could not do so completely, or even to a very large extent, because of the differences in the manageability of the two media. Even after sound was added to film, increasing the imitation of the real sensory environment, the kind of information which film could be used to relay was different from, and far more limited than, the kind of information which could be relayed by print. Also, the ease with which the information could be recovered 'on demand' was very different. In some ways, the development of cinema paralleled the development of print (it gave birth, for instance, to a whole new kind of artistic expression), but the priorities had to be different. Print remained the most convenient storehouse of knowledge and ideas, while cinema became paramount as a purveyor of certain kinds of fantasy.

The second stage of the revolution was not, however, quite complete. In the cinema, sound was added to the medium of film. In the meantime, thanks to television, pictures were added to the auditory medium of radio.

At first glance it is not easy to see that there is a crucial difference between cinema and TV – after all, both are sound plus pictures even if in the one case A was added to B and in the other B was added to A, and TV shows old cinema films in considerable quantity (there is a generation familiar with cinema and cinema history almost entirely thanks to TV). But there *is* a crucial difference.

TV is an extension of radio, and it retains – as cinema does not – radio's property of instantaneity. It is *not* primarily a preserve-store-and-distribute medium like cinema and print, although it can be used to present such preserved-and-stored material. Primarily, it is a destroyer of distance, giving the viewer access to things which are happening in front of the camera – extending the perception of sight just as radio extended the perception of hearing.

TV requires active participation on the part of the viewer, but (unlike print) it does not require the switching-over of sensory orientation and (unlike cinema) it does not require removal to an artificial environment. When a viewer switches on a TV set he does not abstract himself from the real world, but brings a new source of information into it. Its information-presentation involves the same sensory orientation which is adapted to the perception of that environment. Thus, people can watch TV while eating dinner, ironing shirts, making love or compiling lists (but not while reading).

Television represents the culmination of the second stage in the communications revolution. It has the auditory dimension of simultaneity and the visual dimension of experience-simulation. While the addition of sound (and colour) to cinema only made the picture palace a more complete artificial environment (leading naturally to such experiments as 3-D and quivering chairs for earthquake dramas) the conversion of radio into TV allowed more comprehensive information to be transmitted just as quickly. TV can, if the viewer so wishes, make the sitting-room into a tiny picture palace, but it is much more versatile than that – the picture palace is constantly invaded by 'live' news, 'live' sport and 'live' talk. The element of *participation* which is so important in keeping popular radio alive (with requests, phone-ins and gossiping disc-jockeys) is also important in TV – although it is not, at present, exploited quite so fully.

When the visual dimension is added to the telephone and the gramophone (and video cassettes for use with a television set are already in limited use) the full potential of the new range of communication media will be exploitable. Even then, the new media will by no means *replace* the print medium, or even the cinema medium (which perhaps seems more threatened – by the TV), for these media have unique attributes which will continue to be exploited. The monopoly of print has, however, already been eroded.

The effect of the print monopoly on information-handling was that the information-environment and the experience-environment were totally separated. The totality of that separation is no more. Television recombines the information-environment and the experience-environment, putting the one back into the other, and cannot help but act in opposition to the attitudes cultivated by print. The polarisation of the mind, with the development of distinct categories of 'learning' and 'living' so characteristic of nineteenth-century man, is no longer possible today. In a sense, the intellectual 'wholeness' of pre-print (perhaps pre-writing) man is being restored by the new media, at least to some extent.

There will always be sections of the community who find the unique properties of print necessary to their patterns of informational intake (and in developed countries these sections may never again constitute a minority) but the effect of print on human nature and human ideas can never again be so profound.

How, then, is man being re-made by his marriage to the machine? He is re-defining work and escaping from some entrenched habits of thought. Is that all we can say? If we are going to adopt a general viewpoint, all we can do is deal in general trends. These are the *kinds* of change which are being forced upon man by marriage to the machine. Specific changes, more detailed accounts of change, can only be worked out with reference to specific groups of people, whose relationship with specific groups of machines is known in more detail. Rather than extending this discussion in that direction, though, I think it might be more interesting to look at another aspect of the question of the marriage of man and machine: how long can it last, and how will it change in order to endure?

There is no doubt that the marriage is not nearly so happy as matchmakers like Bacon and Mercier claimed that it would be. The twentieth century has thrown up a number of severe critics who claim that no reconciliation is possible and that instant divorce is the only way to avert domestic tragedy. Even the harshest critics of the machine admit that man and the machine must continue to associate socially, and even live together, but the present institution of marriage, they claim, is a prescription for catastrophe.

The arguments of the would-be marriage-breakers are simple and powerful. First, they point out that both we and the machines are faced with the prospect of imminent starvation. Second, they point out that the entire biosphere – including man – is threatened with being poisoned by the excreta of the machines, and even by the produce of the machines.

The first claim, in both its aspects, is based on Malthusian logic. Populations (unchecked by plague, war or predation) increase exponentially – i.e. *ever faster* – and thus use up their resources ever faster. The production of those resources can, however, only increase linearly. Inevitably, there must come a point at which resource-production is overtaken by resource-demand. In the case of man, resource-production means food-production. In the case of machines – which, despite the advent of atomic power are still almost wholly dependent upon fire power – it means the discovery and exploitation of coal and oil.

The Malthusian principle is a drastic oversimplification of what actually happens, but it is, in essence, inescapable. No matter how fast resource-production increases, the population increase of both men and machines has the ability to overtake it. The implications of the Malthusian principle are by no means so simple. Much argument has been devoted to predicting the magnitude of the catastrophe and its timing. But what kind of a catastrophe will it be?

The popular conception of the Malthusian trap is that once the population explosion reaches a certain point millions of people will starve to death. It is as though we were effective wolves rather than efficient ones, eating up all the sheep and then finding that there was nothing left to eat. This actually happens to some populations of micro-organisms – particularly experimental populations set up in the laboratory. It does not, however, happen very often to wolves.

Population size is subject to limiting factors. Population increase cannot go on indefinitely, or even as far as mathematical extrapolation can take it. If the limiting factors of disease, war and predation are minimised then, as Malthus pointed out, some other limiting factor will come into play – food supply being an obvious candidate. In actual fact, the great majority of the world's population already live in circumstances where the limiting factor on population growth is food supply. Millions of people are starving today, were starving a century ago and a millennium ago, and will still be starving tomorrow and the next day. Famine relief helps individuals but does nothing to alter the situation; the population simply expands to its new limits. The issue at stake is not whether population growth is to be controlled or not, but *how*. We have, in theory, a choice between birth control and starvation control. The great majority choose the latter, for a whole host of perfectly good and practical reasons. The argument about the timing and magnitude of the catastrophe is really a false argument – the problem is not how many or who are going to die when (everybody dies somehow at sometime) but how we are going to live in the continuing situation. It is the tendency among modern catastrophists to concentrate entirely on those who are going to die and not to bother thinking about those who are going to continue living. It is all very well to say that the extrapolated population of the world in the year 2000 (or whenever) is ten billion (or however many), six billion of whom cannot be fed and will therefore have died, and had thus better not be born at all. There will still be four billion (or however many) living, and perhaps their problems need a little thinking about too.

The same argument, essentially, applies to machines and natural resources. The population of machines is subject to limitation, by fuel if by nothing else. Again, the choice is fairly straightforward: control the machine population strategically or allow it to be controlled 'naturally'. Again, and for a host of good and practical reasons, the great majority choose the latter. At some time in the future, therefore, the machine explosion will be curtailed. Again, perhaps we should be more interested in the survivors than the non-survivors.

It is an inescapable truth that we cannot maintain our present lifestyle indefinitely, and that the direction of change which we are trying to pursue cannot be pursued very far. There has, however, been no period in history when this has not been true. Circumstances have always modified the scale and direction of change, and no life-style has been able to sustain itself indefinitely despite the most strenuous efforts of its adherents. In that sense, we live in the midst of a continuing catastrophe, and are perpetually faced with the problems of how to live with it and how to avoid dying of it.

The second argument of the marriage-breakers – that the world is in danger of being poisoned by the machines – is complementary to the first. Again, some micro-organisms – especially those living in artificial conditions – are poisoned by the build-up of their own excreta (the standard example is yeast in alcohol). Again, this 'proves' little more than the fact that there is no escape from a test tube.

Essentially, the pollution argument is that the machines themselves are becoming a limiting factor affecting human populations, either directly or – by threatening ecological balance and food supply – indirectly. The notion of catastrophe is perhaps more appropriate here; ecological balance is in many instances a delicate and unstable thing, and the removal of one element from a complex system of interrelationships can have quite far-reaching effects on the system as a whole. The population-regulations of sheep parasites is a matter of some importance to the wolves. We are still, however, dealing with limiting factors and not with an either/or situation. We cannot go on very long as we are going on, but it is almost inevitable that we should choose to go on as we are *while we can*, until circumstances force us to change our minds.

The arguments of the marriage-breakers are generally couched in quantitative terms – usually in quantities of death. It is, they say, impossible for us to go on, we should recognise that fact, and we should start changing now. There is, however, more than one kind of impossible. Politics, as some cynic once remarked, is the art of the attainable, and 'the attainable' almost invariably means what can be done or got tomorrow, not providing for the indeterminate future.

There are, however, qualitative arguments against the marriage of man and machine as well. There are a great many people who claim that the quality of life within the marriage is inferior to the quality of life without. Many also set out to prove it by adopting alternative lifestyles. Curiously enough, a great many of those who choose to stay with the marriage believe, or at least sympathise with, those who do not. (The same kind of doublethink seems to apply in very large measure to interhuman marriages as well.)

There seems to be little doubt that the majority of people involved in the marriage to the machine believe that it is worth sustaining. They either believe that it can be adapted to our needs or that even if it cannot it is worth hanging on to until it is finally driven on to the rocks. (Again, the same thinking generally applies with respect to human marriages.) However, the idea that in crowding together in great cities ('concrete jungles' or 'human zoos') we are doing terrible things to ourselves – or that, at least, we have no real idea of *what* we are doing to ourselves – is one that is gaining currency. Increasingly, dissatisfaction and misery and aberrant behaviour (all, occasionally, under their collective *alias*, mental illness) are blamed on the modern style of life. The machines, like all worthy spouses, support us in the manner to which we would like to become accustomed, but – inevitably – overt glamour and ease conceal covert problems and discomforts.

As every marriage counsellor believes, we are hardly capable of perceiving our own situation, let alone of managing it. Having extended the concept of sickness to the mind as well as to the body, it seems a natural analogical step to extend it to society as well. Environmental stress is taking its place alongside bacteria and viruses as one of the major causes of physical malfunction, and this introduces us to a new class of effects which the man–machine marriage – or the lifestyle in which it is conducted – may be having on its participants.

Modern catastrophists have become enamoured of extracting 'lessons' about the hard facts of nature from examining what befalls animal populations in highly unnatural circumstances. The effects of crowding in modern cities is compared to the effects of crowding in animal populations – particularly rats. Calhoun's experiments with rat cities, inducing 'population explosions' and studying the precise pathology of the carnage which follows, have been presented by several prophets as moral tales.

The first study of an animal population explosion was conducted by John Christian in the 1950s, when a population of Sika deer introduced to a small island in Chesapeake Bay underwent a population crash. Christian performed autopsies on many of the dead deer and found that in almost all instances the adrenal glands were greatly enlarged and abnormal in structure. The adrenal glands produce adrenalin – a hormone released in order to prepare the body for fast action in times of stress by increasing the heartbeat and the metabolic processes making energy available to the muscles. Usually, it is produced in quick bursts and disappears after the threat to the organism is alleviated. The deer examined by Christian, however, had apparently fallen victim to a 'stress syndrome' in which continual stimulation of the adrenal glands had resulted in the constant adrenalinisation of the animals, leading to undue strain on accelerated physiological systems. The deer had become 'highly strung', had suffered physiological deterioration, and their overstimulated hearts had been rendered all too likely to fail completely as a result of sudden shocks.

Calhoun, following up Christian's findings, discovered precisely the same kind of physiological change occurring in rat populations allowed to increase by food supply but confined in a limited space. The physiological symptoms had behavioural effects, too. The rats became 'neurotic', becoming aggressive or cannibalistic. Maternal behaviour was disturbed, and generally speaking, such 'social system' as rats might be said to possess collapsed.

The stress syndrome was discovered in nature and accounted for two long-recognised but ill-understood enigmas – the cyclic changes in small mammal populations in North America and the curious periodic behaviour of lemming populations in North Europe and Russia. The North American cycle was attributed to the disturbance of the ecological balance by population fluctuations in the snowshoe hare, which − every few years − undergoes a population explosion followed by a population crash. The same thing happens to the lemming. In each case, the population crash is brought about by the incidence of a stress syndrome, when perpetually adrenalinised individuals begin to exhibit aberrant behaviour which almost invariably results in massive death-rates. The snowshoe hares tend to keel over and die in convulsive spasms, the lemmings − not the best-tempered of creatures at any time − become very fierce and disturbed. They tend to migrate westward, ultimately − if they do not get killed *en route* − continuing into the sea where they drown.

Enough snowshoe hares and lemmings always survive to begin the cycle all over again, but there is another aspect to the stress syndrome apart from its effect on the mortality rate, and that is its heritability. The syndrome is not, of course, communicated genetically, but its effect is at a maximum during the breeding season, and the foetus in the womb shares the adrenalinised blood of the mother, thus being born in a state of adrenalin shock. The viability and probable breeding success of such a foetus is affected quite markedly by this. The stress syndrome thus destroys a percentage of one generation and reduces the viability of the next (though it only takes another six or seven years for the population to reach crisis point yet again).

It must be emphasised that the stress syndrome is *not* an 'evolutionary adaptation' developed by these species to 'control' their populations. It is an accident of nature − like cancer, one of the 'thousand natural shocks' that flesh is heir to because that is the way bodies are built. Is there, then, any logical reason why we should not expect the syndrome to become manifest in human beings under conditions of extreme crowding and great stress?

The obvious answer is no. Indeed, the physiological aspects of the stress syndrome are quite obvious in modern life, especially in big cities. Ulceration of the gut, high blood pressure and heart disease are the commonest causes of serious illness, and heart failure due to stress is by far and away the commonest cause of people dying 'before their time'. The widespread use of ataractics to combat stress and tension has already been pointed out. It has been alleged by some psychologists that there is a clear correlation between the incidence of psychosis and the density at which people live. Attempts have also been made to 'explain' criminal behaviour according to the same pattern.

All this evidence suggests quite strongly that the stress syndrome may have a considerable part to play in human population limitation during the next hundred years. The most threatening aspect of the suggestion is the fact that we are, so far, only in phase one. We are not yet able to appreciate fully the second-generation effects of shock-diseased children.

The source of stress in human society is not simply linked to crowding, but to the demands put upon the individual in coping with the complexity and frustration of life. Stress can be offset by making use of adrenalin in periodic bouts of strenuous activity: participation in manual labour or in sport, particularly. Though we have had concentration-cities for some time, it is only in the last twenty-five to fifty years that our lifestyles have changed so that most of the stress we encounter is encountered in sedentary situations, and so that we give ourselves less and less chance to make use of the adrenalin in activity.

These are things worth thinking about seriously. However, the perspective we obtain from the animal experiments and situations so far discussed is rather one-sided. It misses on important aspect of the problem.

Calhoun worked with mouse populations as well as rats, and so did Alexander Kessler, who attempted to follow up this kind of study still further. Both Calhoun and Kessler found that most of their mouse populations followed the same kind of pattern as the rat populations – stress syndrome leading to population crash. Both, however, also found that a small number of mouse populations did *not* fall victim to the syndrome, but continued to increase to quite incredible densities (of the order of 100 per square foot – literally standing room only) without showing the symptoms of constant adrenalinisation. These populations contrived to adapt – not genetically, but *behaviourally* – to the situation. They abandoned the rigid 'pecking-order' social hierarchy, and did not become particularly over-aggressive. The infant mortality rate remained high, and there was a degree of failure in maternal behaviour, but in general the kind of social disintegration and population crash which happened invariably to the rats did not occur – the population tended to stabilise at its upper limit.

This puts a slightly different complexion on the problem. Man, after all, is the most behaviourally versatile of all the mammals, and one would expect that if mice can adapt to crowded conditions, then men ought to be able to do so too. What we might expect to see, if such behavioural adaptation were to take place, would be the breakdown of social institutions based on pecking-order systems and territorial aggression, and their gradual replacement with systems providing situations with less inherent conflict and tension. Undoubtedly, the transition would be very rough, and one would expect to see both the ratlike stress-induced neurosis and the mouselike stress-conditioned behavioural adaptation going on at the same time. How far contemporary stresses on our social system and the processes of change within it may be understood in the light of an analogy such as this is, of course, open to question, but it may provide a useful perspective for looking at modern cultural evolution.

Emile Durkheim, one of the first men to attempt a standpoint of scientific objectivity for the study of society, wrote a classic study of suicide in 1897. In this book he developed the concept of *anomie* (French for 'lawlessness') to describe a sense of isolation which overtakes many city-dwellers as they feel that community life and values are being eroded and becoming meaningless. This concept is, perhaps, becoming much more widely applicable in today's world.

Today, we often hear the complaint that people are losing their individuality. Resentment is expressed when people feel that they are 'being treated like statistics'. There was a time when a man who went into a shop was a 'customer', but nowadays he is a 'consumer', adapted to the use of standardised goods in standardised packaging, the puppet of the advertising industry. The products of modern industry are disposable, and the relationships between modern people have become caught up in the same regime of transience, and are disposable too. The business of living has become enmeshed in bureaucratic semi-order; we are minds living within a greater 'mind' whose information-handling and capacity to organise action seem to us to be plagued by an incredible stupidity.

The instantaneity of communications is perhaps a two-edged sword. On the one hand, it permits us easier access to all the information we want and need. On the other hand, it makes us vulnerable to communication which we do not want and do not need; the constant assault upon our senses and emotions by the 'hidden persuaders' who seek to manipulate our lives – physical and mental – according to one set of precepts or another. The new media have paved a golden road of good intentions for the artificial mythologies of advertising and propaganda. The complexity and diversity of the information-input to our minds today is far greater than it has ever been before, and it becomes more and more difficult to exercise judgment with respect to this bewildering array of dishonest data. Learning to meet and cope with the demands thus put upon us is a slow and uncertain business.

It is becoming very much more difficult to create 'personal space' from which one's intercourse with the world may be negotiated and regulated. The establishment of such space publicly and territorially ('an Englishman's home is his castle') is no longer so easy when the home is subject to constant 'information-invasion' via the new media. Increasingly, personal space has to be re-created internally and privately, inside the mind. Small wonder that feelings of isolation and loneliness are prevalent in cities, where the information-environment is at its most intense.

Further strain upon the mind is generated by virtue of the fact that the information-environment is not only more complex but changes ever more rapidly. Information dates very quickly, and the persuasive information with which we are bombarded changes most rapidly of all. If human beings were really dependent on conditioned reflexes to determine their actions the system would have cracked up long ago – the rate at which yesterday's myths are torn down and new multi-story myths erected in their place is quite tremendous. Perhaps the most eloquent testimony of this is the rate at which our languages change. Of the half million or so English words liable to be used today only half would have been meaningful to Shakespeare, and with respect to the common words used in everyday conversation the proportion would be just as small if not smaller. The 'turnover' in fashionable slang is very considerable – the 'lifetime' of a popular slang term may be measured in months rather than in years.

Within these generalisations about modern life can be found the particular sources of human stress. Tranquillisers, ulcers and coronary thromboses notwithstanding, we are apparently bearing up very well and adapting to the new information-environment with a fair degree of success. We must not, however, lose sight of the fact that we are being called upon to use every last vestige of our considerable behavioural adaptability and versatility.

Living in the world of the man—machine marriage is not easy — certainly not as easy as the technological Utopians anticipated. And we must also remember that change will, in the future, tend to accelerate rather than slow down as new limitations come into force and bring with them new challenges to ingenuity and adaptability.

We have become cynical about the miracles of the machines (and, for that matter, about science itself). We no longer see the marriage of man and machine as a ticket to happiness, but as an inordinately complex and difficult relationship. There is, however, very little chance of a divorce, or even a relaxation of intimacy. One thing, though, is certain: the honeymoon is over.

The Question of Scientific Belief

Science is a particular brand of disciplined knowledge. It is not by any means the only brand available. The fundamental assumption of science is that the universe is a systematic, ordered and rational place whose principles are amenable to analysis and comprehension. Ideally, scientific knowledge should begin as hypothesis and be rigorously tested by experiment so as to exclude all alternative hypotheses, establishing the correct one by failing to falsify it. In practice, it is not always possible to do this.

Most major religions present their own particular brands of disciplined knowledge, too. The fundamental assumption of religious thought tends to be very much the same as that of science − that the universe is a systematic, ordered and rational place whose principles are amenable to a limited degree of analysis but not necessarily comprehension ('God moves in mysterious ways' and 'there are things which man was not meant to know'). As a fundamental assumption, this is perhaps more reasonable than the assumption of science, in that it recognises human limitations. On the other hand, it is not such a useful assumption because it does not encourage exploration of those limitations.

The real difference between religion and science, though, tends to be the method by which knowledge is accumulated and systematised. While new scientific knowledge is revealed by observation and experiment (which can, supposedly, be carried out by any man, though organisation into systems is usually the prerogative of special men involved with the quality of 'genius') new religious knowledge is revealed by divine enlightenment (which may happen to any man, though organisation into systems is usually the prerogative of special men imbued with the quality of 'holiness').

Science thus claims fidelity to the truth of analytical logic and repeatable observation. In answer to the question 'How do you know this is true?' the man of science may reply along the lines: '*I* saw it happen, *he* saw it happen, and if you do the experiment correctly *you* will see it happen too.' Religion claims fidelity to a different kind of truth – the truth of divine revelation and repeatable affirmation. In answer to the same question the man of religion might reply to the effect that: 'God told *me* so, God told *him* so, and if you listen properly, God will tell *you* so too.'

Both religion and science tend to base their propaganda – their 'sales pitch' – on grounds which are mainly pragmatic. Think our way, they say, and you'll find that it works. Live by these rules and they won't let you down. Both science and religion can provide a well-scripted narrative history which testifies to great success, and each can supply a full and horrific account of the other's failures and disasters.

Both religious orthodoxy and scientific orthodoxy tend to be outraged by unbelief. They express this resentment (or horror) by labelling unbelievers with terms intended to vilify. The Christian religion has evolved such cutting word-weapons as 'heretic' and 'pagan', and has contrived to inject venom into purely descriptive terms like 'atheist' and 'Jew'. Science has invented the 'crackpot' and the 'crank', and made terms of abuse out of purely descriptive words like 'dogma' and 'mystic'. Neither religious orthodoxy nor scientific orthodoxy has hesitated to call its most voluble critics 'fanatics' or 'lunatics' (in fact, the accusation of insanity has served them both very well). During periods of history when it enjoyed considerable political power the Christian Church has imprisoned, tortured and executed its opponents with prodigal liberality, and one of the perennial nightmares of modern machine-critics is that if the scientific establishment ever achieved political power of the same kind similar cruelties might be perpetrated with a similar lack of conscience (as, indeed, they were in Nazi Germany).

Today, science has the upper hand in education. It was not always so, and science represents the achievement as the victory of rationality and the triumph of truth. Certainly, it is the former, but does the former necessarily imply the latter?

One of the most remarkable things about beliefs which prosper in modern highly technological societies is the persistence of belief in brands of disciplined knowledge which are non-scientific. It is not the mere existence of alternatives to science which is surprising so much as their extreme diversity. There are far more brands of belief being hawked around in bookshops than there are brands of soap powder on offer in supermarkets – perhaps ten or even a hundred times as many. A great many of these brands of belief are of recent origin, and are attacked by scientific orthodoxy and religious orthodoxy alike. Some mimic the packaging of science, some the packaging of religion, but all attract adherents. Why?

The simple fact is that neither science nor any other brand of disciplined knowledge is providing a product satisfactory to all its potential customers. Resentment of this fact is probably far stronger in modern science than in the modern Church, which has learned a degree of tolerance. The resentment of science is perhaps particularly strong because it sees itself as dealing in incontrovertible truth, with a monopoly on rationality. It is particularly bitter in its attacks upon those alternative brands which pretend to the trappings of science – observation, evidence, experiment, and so on, but which use them as a basis for the support of what the scientist may believe to be irrational conclusions.

In this respect scientists often behave as though there were something sacred about the scientific method, as if science were a religion which could tolerate no heresy. In this chapter I want to look, in a fairly general way, at some of the alternatives to the brand of knowledge which science is marketing. The principal questions I want to keep in mind are: what do they have in common with science and in what way do they oppose it, and, why do people often choose to believe in them rather than in the 'accepted orthodoxy'?

I feel that I should make my own position clear. Far too many investigations of this sort are conducted from an absolutely committed standpoint which pretends to be objective and neutral. I think it would be dishonest to pretend to be sitting on the fence between orthodoxy and unorthodoxy when I am not. I am a sceptic. I do not believe in astrology, phrenology, theosophy, chiromancy, graphology, numerology, pyramidology, scientology, psychic surgery, telepathy, psychokinesis, levitation, psychoanalysis, vegetarianism, flying saucers, ghosts, goblins, astral bodies, ectoplasm, perpetual motion, dowsing, Biblical extraterrestrials, the hollow Earth, the World Ice Theory, orgone, the Loch Ness monster, Atlantis, the prophecies of Nostradamus, Ossian, Adam and Eve, Superman, systems for winning at roulette, True Confessions, the integrity of the advertising profession, the impartiality of the law, the public spirit of politicians, Noah's ark, monkey glands or the sanctity of womanhood. However, I am prepared to extend my scepticism – as many fake sceptics are not – to the theory of relativity, the theory of evolution by natural selection, and the Cartesian principle. These things seem to me to be more or less reasonable, and I can find no reason to refuse to accept them as workable ideas, but I also see no need to commit myself to believing in them.

I think someone once defined faith as 'believing in something you know damn well ain't so'. Lack of faith, I presume, involves being prepared *not* to believe in the things 'you know damn well *are* so'. When we begin to believe in the things we know then we give away the ability to modify our knowledge. We abandon our inalienable right to be wrong. All discovery comes out of doubt, and once we begin to believe something, and to have faith in it, we surrender doubt and it becomes irrelevant whether our knowledge is true or not.

Voltaire is often credited with the assertion that even if he did not agree with what someone was saying he would defend to the death their right to say it. Most people adhere to a slightly different principle – that they may not know what they are talking about but they will defend to the death whatever it is they happen to be saying. Voltaire believed in nothing, most people believe in themselves.

Belief is apparently necessary to a great many people, because virtually everyone has it. They believe in widely different things, but they all have in common the single belief that they are right and others wrong. But why is belief necessary? Can we not say that one theory is more likely than another without saying that one is certain and the other impossible? Can we not accept, merely for the purposes of deciding what action to take, that a particular principle is *useful* without making that principle a foundation-stone of universal logic which everyone else must recognise, and to which there can be no conceivable exception?

Every one of the things I do not believe in is supported by evidence of some kind. The fact that I do not believe in them does not lead me – as it inevitably does committed believers – to say that all that evidence is either fake or worthless. The committed believer can only be wrong, for even if all the things he believes in do happen to be true they would be just as true if he were prepared to doubt them while accepting them as rules of thumb. He can never be 'more right' simply because he commits his belief. We should, I think, be prepared to recognise that all evidence is relative, and though some ideas may be supported by 'more' evidence or 'clearer' evidence there can never be any ultimate certainty – not even *Cogito ergo sum*.

This is a statement made by a man of science (T. R. Willis) about extra-sensory perception and its adherents:

The conclusions of modern science are reached by strict logical proof, based on the cumulative proof of numerous *ad hoc* observations and experiments reported in reputable scientific journals and confirmed by other scientific investigators: then and only then can they be regarded as certain and decisively demonstrated. Once they have been finally established, any conjecture that conflicts with them, as all forms of so-called 'extra-sensory perception' plainly must, can confidently be dismissed without more ado.

This is a barefaced lie. This is not a true picture of the way the conclusions of modern science have been reached (as I have tried to make clear in this book), and even if it were those conclusions would not be regarded as 'certain' – merely as very probable. The idea that any conjecture conflicting with scientific orthodoxy must be dismissed out of hand (with all its adherents labelled 'crackpots' and all its evidence labelled 'fake' and 'worthless') is a corollary of the most committed belief. Personally, I find the evidence for ESP unconvincing. But I do not feel that I am thus entitled to dismiss that evidence as worthless or fake.

The picture of scientific knowledge which Willis is presenting is a kind of glorious ideal – if only all our knowledge *could* be so wonderfully confirmed and rigorously tested. Alas, for the most part it cannot. There are limitations to the kinds of test we can carry out. We have only one tenth of the pieces of the jigsaw, and to fill in the full picture we have to use imagination. It is hypocritical for scientists to lay claim to any kind of perfection for scientific knowledge.

Science is not a religion, but there are a great many scientists who behave like high priests. It seems that men of science need beliefs no less than other men, and make their science into a kind of personal religion, with its dogmas and its heresies. The image of the scientist as an objective seeker of truth, though strenuously promoted by scientists, is in fact a false picture. Most scientists know what they are looking for before they go out to look, and they use the eye of faith in making their observations. Einstein, Dirac, Darwin and many others built their discoveries on a pedestal of faith. Their triumph was not the ability to doubt everything, but merely the ability to doubt what had gone before.

Charles Fort made a lifetime's hobby out of collecting pieces of information 'damned' by science – data which were dismissed as unreal and impossible, largely consisting of eye-witness reports of strange phenomena published in newspapers. He published several books in which he organised these data into categories and pointed out some of the conclusions which might be drawn from them – such conclusions as the fact that 'we are property'. Fort refused to believe in any of the theories of scientific orthodoxy, but he also refused to believe in any of the theories he had himself formulated from the damned data. He refused to join the Fortean society which was formed to promulgate his ideas. Fort's attitude is, I think, a reasonable one; he was ready to admit that he did not understand the workings of the universe and was content in his ignorance. He was ready to mock all pretensions on the part of others, and his books are the expression of that mockery – they are conscious mimicry of science, gathering data and forming theories to account for them. The chief difference between Fort and the orthodox scientists was that they damned *his* data while he damned *theirs*. His work showed up the weakness of the priestly ambitions of the scientist (and it is rather a pity that the Forteans made his work the basis of a faith of their own).

Alfred Jarry, the French pioneer of the 'theatre of the absurd', created, in reaction against nineteenth-century rationalism, the science of *pataphysics*. While other sciences attempted to draw generalisations from assemblies of particular data, pataphysics was to draw the particular out of concepts of the general. While physics dealt with rules, pataphysics was to deal with exceptions. This, too, is a mockery of the faith of the scientists – the faith that science has the one and only truth, and has it all wrapped up and ready to go.

In so far as scientists have attempted to make science into a replacement for religious belief they have failed. Their wares are not attractive enough, the beliefs which they sell are not useful in the way that the customer wants to use them. It is for this reason that so many other sellers of belief have tried to co-opt the veneer of scientific methodology – to borrow the weaponry of the priests of science – but have used it to serve up goods with much more consumer-appeal.

It is noticeable that a great many of the scientific 'heretics' whose work has attracted adherents in the last few decades have done nothing more than combine the modern mythology of science with the older mythology of religion.

Immanuel Velikovsky, for instance, has put forward in three commercially successful books (*Worlds in Collision*, *Ages in Chaos* and *Earth in Upheaval*) the idea that between 1500 and 700 BC the solar system was upset by a series of disasters. Jupiter is said to have collided with Saturn, and a fragment knocked out of the larger planet then interfered drastically with Earth and Mars before bits of it became the asteroid belt and the rest of it became the planet Venus.

There is nothing particularly unusual about these ideas – they are simply old-fashioned catastrophism applied in a slightly naïve manner. Like the eighteenth- and nineteenth-century catastrophists Velikovsky attempted to explain the changes which the Earth's surface had undergone by reference to a very small time-scale – a matter of centuries rather than hundreds of millennia. Like his historical antecedents Velikovsky turned for his evidence to the Bible, and made the Biblical miracles – the flood, the plagues of Egypt and the parting of the Red Sea etc. – into side-effects of the cosmic disaster.

Where Velikovsky differed from the old champions of flood and divine intervention was in the meticulous and assiduous nature of his search for evidence. He was concerned not only to recruit evidence from the Bible, but to *explain*, by means of his theory, every last supernatural event therein. When he had finished with the Bible he went on to the mythologies of other religions, 'explaining' the apparently inexplicable wherever he found it.

His basic philosophy is that of the scientific faith: behind the chaos of singular events there lies a network of unchanging, comprehensible principles. There is, however, a strong Fortean streak in him: he is cynical about a great many observations made by orthodox scientists and about the conclusions drawn therefrom, and at the same time he is a diehard champion of 'damned data' – data which orthodox scientists do not admit as real.

Velikovsky's arguments are not so very unscientific, although they often raise the ire of scientists because they conflict with data which are generally held to be true and accurate while leaning heavily on data which are generally held to be worthless. The consequence of the way that Velikovsky selects his data, however, is that the aesthetic appeal of his theories is very different from the aesthetic appeal of such abstract theorists as Einstein and Dirac. It is the aesthetic considerations which make Velikovsky's ideas so popular relative to those of orthodox science. His concepts are a little more simple than those used by modern physics, but the essential advantages they have is that they are so much easier to visualise mentally and come to terms with. They appeal to the deceptive logic of common sense. The imagination is seriously challenged by the time-scale of evolution and cosmic events, and it is really quite a heroic task to attempt intellectual mastery of Einstein's theories and the twentieth-century model of the atom. Velikovsky, essentially, dismisses modern science as so much mystical rubbish, and restores the comfort of common sense to the business of understanding why the world is the way it is.

At first, Velikovsky's heavy reliance upon the Bible as a source of evidence makes it appear as if he were defending the Christian religion against the ravages of scientific logic, but this is not so. Velikovsky takes for granted not the validity of religious thinking but the validity of the *sacred writing*. I commented in an earlier chapter on the importance of writing in the founding of organised religion, and here we can obtain a significant insight into the fact that it is the writing, and not the thinking behind it, which is vitally important to the sustaining of religious belief. When Darwinism threatened religious faith the primary reaction of many committed Christians was to try to defend the *literal* truth of the Bible. The fundamentalists, who are still waging war upon Darwinism in several American States (having failed in the attempt to have Darwinian theory banned from schools they are currently campaigning to have the Book of Genesis given equal weight as an alternative scientific theory), have always been concerned with the absolute veracity of the Biblical word, and would never admit to the possibility of any metaphorical or allegorical interpretation. It seems, in fact, that a good many people find it far easier to believe in a reified and trivialised God (a cosmic disaster or an astronaut) than a sacred book which uses all the potential of language and myth instead of absolute truth and sober reportage. After all, myth and allegory, like modern science, trespass beyond the limits of common sense.

The most prominent – at least, the most financially successful – of Velikovsky's followers is Erich von Daniken, who uses similar methodology and philosophy to support a different theory. In von Daniken's view all the damned data of science point to the fact that the Earth has been constantly visited by and interfered with by extraterrestrial visitors who have casually steered the course of human history. Von Daniken's theory has one advantage that Velikovsky's did not – it can be used not only to explain 'supernatural' events, but also to explain thousands of human artefacts and works of art whose significance and purpose now seems somewhat enigmatic. In von Daniken's view, not only do supposedly sacred writings show no trace of imagination, linguistic sophistication or myth, but no human being who lived more than two thousand years ago was possessed of any imagination, ambition or spiritual aspiration which might be expressed in the things he made or built.

Like Velikovsky, von Daniken carefully mimics the scientific method – he selects his data, then shows how they can all be explained by a single hypothesis, and argues that no other hypothesis could explain this particular selection of data. As is the case with orthodox science, his basic assumptions are aesthetic ones. Like Velikovsky, he provides a programme for understanding which is moderately simple and quite easy to grasp, always appealing to common sense.

Velikovsky and von Daniken both make capital out of a curious kind of martyrdom. Orthodox science persecutes them both, vilifying them and labelling them cranks. Thus they exploit the tendency which many scientists have to treat science as a religion. *Chariots of the Gods*, von Daniken's most famous work, begins with the words 'It took courage to write this book and it will take courage to read it'. This is a most seductive lie, implying as it does that belief in the theories the book contains will expose the reader to a persecution which is intolerable to any liberal and free-thinking man. It has always seemed to me that it takes far less courage to reject out of hand the difficult ideas and theories of modern science than it does to try to understand their meaning and implications. To believe in a world where common sense is an absolute ruler and all enigmas have the same comforting explanation is, I think, to take the road of least intellectual resistance.

Velikovsky and von Daniken do provide genuine alternatives to scientific orthodoxy. If one is going to commit one's belief at all, there is no reason why one should not believe in their theories rather than the beliefs marketed by orthodox science – belief has nothing to do with truth, and beliefs are chosen by virtue of aesthetic considerations. Only if one is prepared *not* to believe – to doubt everything – can one actually decide whether one theory is more likely than another, because only then can one compare the relative likelihood of the data instead of selecting the favourable ones and damning the rest.

Velikovsky and von Daniken not only show up the fact that the priests of science are marketing an unpalatable product, but they also show up – by parody – the questionable nature of the methods by which the priests of science arrive at their unpalatable product. The 'facts' recognised by scientists are really no different in kind from the 'facts' recognised by these alternatives to science, although scientists claim that they are. There is no such thing as an empirical fact – an observation without a context. We see by interpretation, and what we see depends on how we interpret. In practice, the theories we believe reflect the kind of universe we want to live in – it is the theories which we entertain without believing which may ultimately allow us to approach an understanding of the universe we *do* live in.

The scientific revolution which got under way after the Renaissance was aided by a new philosophy – the idea that a better knowledge of God might be achieved by the dedicated study of the universe He had made. Scientific inquiry was blessed by the Church wholeheartedly, on the understanding that what the scientists found would confirm and lend more credence to the things which the Churchmen already believed. In point of fact, the observations of the scientists did *not* confirm what the Churchmen believed, and as science has progressed the measure of agreement between scientific observation and religious faith has grown less and less. But there has, understandably, always been a considerable demand for a brand of 'scientific' truth which *would* confirm what people did believe and wanted to go on believing. The reason the Church of the seventeenth century was so determined to uphold the Aristotelian theory of astronomy against the Copernican model was that it had been demonstrated by Thomas Aquinas that there was no point of contradiction between Aristotelian science and Christian dogma – it was a science which confirmed religious belief. That – from the point of view of the Church – was what science was supposed to do. That was what it was *for*.

This kind of thinking has never become extinct. There have been innumerable attempts to reconcile new scientific knowledge with religious beliefs and, conversely, innumerable attempts to show that old beliefs really had a perfectly good scientific basis all along. One successful branch of the Christian faith has represented Christ as a scientist, and all kinds of ancient and arcane disciplines – notably astrology – are today seeking a logic to their rituals which conforms to the aesthetic philosophy of science. In parallel with von Daniken's attempts to explain ancient artefacts by reference to extraterrestrials there is a growing movement to make them into ancient observatories and calculating devices. The cult of scientology began as a new and unorthodox psychiatry called dianetics, but gradually transformed itself into a religion as it co-opted more and more beliefs with long histories of consumer-appeal.

Perhaps the most interesting of all the attempts to provide an ancient belief with a new scientific packaging is the one which provoked the attack I quoted earlier: the attempt to discover by scientific methods the existence of extra-sensory perception.

The case of ESP research is particularly interesting because most of the data on which it is based, though no less 'damned' by scientific orthodoxy than the data used by other proponents of alternatives to science, are largely experimental. Many champions of ESP will refer to second-hand evidence for additional corroboration, but the real basis of their case consists of tests actually carried out in the laboratory under (more or less) controlled conditions. It is in this case more than any other that the logic by which orthodox science seeks to expose the relatively unconvincing nature of the unorthodox data can be turned around to expose the lack of certainty in the data of orthodox science.

The foundations of ESP research were laid by J. B. Rhine in the 1930s, in a series of experiments carried out at Duke University. His basic methods involved a pack of twenty-five cards, each marked with one of five symbols: cross, circle, square, star or wavy lines. The cards were shown, one by one, to an 'agent' while a 'subject', who could not see them, attempted to guess what they were.

Rhine's great innovation was not the establishment of basic experimental methods for the study of ESP but the use of statistical methods to sort and evaluate the data. What he searched for was not people who could identify all the cards without seeing them, but people who could guess correctly a number significantly higher than would be expected by pure chance. The word 'significant' is here used in a special sense, and generally refers to a result whose probability of being produced by pure chance is less than one in a hundred. (Some scientists – and some ESP researchers – are content with a significance level of one in twenty, but one in a hundred is more common.)

The importance of statistics in science has increased vastly over the last hundred years. Clerk Maxwell demonstrated that the temperature of a gas was determined statistically, as the average kinetic energy of the molecules, in the late nineteenth century. At about the same time Darwin produced a theory of evolution that was basically a statistical process – the 'fitness' of an individual in the Darwinian model is really a *probability* of its making a contribution to the next generation. Later, of course, Heisenberg demonstrated that the position of an electron was a probability distribution rather than a point in space. The behaviour of gases, the evolutionary future of animal populations and the distribution of the energy levels of the electrons in a great many hydrogen atoms can be calculated with reasonable certainty (far, far greater than 99 per cent, because of the vast numbers of individual cases involved in the whole sample) but individual cases can be misleading. So can small samples – especially biased samples. (Maxwell's demon, for instance, is a hypothetical biased sampler, selecting slow-moving molecules from a hot gas and fast-moving ones from a cool one so that heat passes from the hot gas to the cool.)

We are probably all familiar with the various ways in which accurate statistics can be very misleading through our acquaintance with advertising. We know that when it is alleged that 'Four out of five cat-owners say their cats prefer it' what may actually have happened was that the sampler went out, found four cat-owners who *would* say that and then took their names plus the name of one other cat-owner. We know, too, that when it is alleged that 'tests show that children using it need up to 30 per cent fewer fillings' what the figures might actually have shown is that *one* of the children using the test product needed 30 per cent fewer fillings than *one* of the children using the control, and that the rest of the thousands involved needed anywhere between 30 per cent less and 30 per cent more. The *average* number of fillings needed by the control group may only have been 1 or 2 per cent greater.

We do not expect to see scientists using statistics in quite this manner, but whenever a scientist is committed to his work – on which his whole reputation may depend – there is always the chance that some kind of bias may creep in. It is often taken for granted by scientific workers at all levels that experiments which 'don't work' (i.e. do not produce the expected result) can be written off and forgotten. There are undoubtedly many instances where a scientist has confirmed an experiment by repetition (and thus elevated it to the standing of a 'scientific fact') by doing it eighty times and getting the 'right' result six times.

There is also another way in which statistics can be quoted accurately in order to lie by implication. One of the most widely used of all statistics (after the 'average') is the correlation coefficient. This simply compares two sets of data to see if they share a common pattern. It is in this way that smoking was correlated with lung cancer – it was found that among heavy smokers the incidence of death by lung cancer was significantly higher than the incidence of death by lung cancer in non-smokers. (Again, 'significantly' here means that the probability of the discrepancy occurring by chance was very low.) This experiment is often taken as evidence – or even proof – of the fact that smoking causes lung cancer. In fact, it is not. The fact that the correlation exists may have several interpretations: it may be that lung cancer causes smoking, that smoking causes lung cancer, or that the causes of smoking and lung cancer, though different, are in some way linked. (It may, for instance, be true that there is a high correlation between the number of convictions obtained for driving while under the influence of alcohol over a period of years and the number of priests ordained over the same period. Both figures increase over a period of time. The probability of the appropriate degree of correlation being obtained *by chance alone* is very low – but there is far more than chance operating in the context from which these figures were drawn.) It is worth pointing out that the non-smokers who did not die of lung cancer were none the less dead, and as well as a positive correlation between smoking and lung cancer the figures must have shown a *negative* correlation between smoking and certain other causes of death. One would not infer that the refusal to smoke 'caused' the higher incidence of other modes of dying.

The link between smoking and lung cancer has been well-established by other evidence – tobacco contains a percentage of known and identifiable cancer-causing agents. But without such supporting evidence, the statistical evidence remains ambiguous.

Statistical evidence, then, can be misleading. Even if the

greatest care is taken, however, to infer no more than can actually be deduced from the data, the statistical method remains fallible. The trouble is that if one finds 'significance' in a result whose probability is one in a hundred then in a long series of experiments – perhaps involving several thousand 'runs' – one is certain to be turning up significant results with considerable regularity. The scientist who begins his work with a degree of objectivity may doubt these significant results but even he is not likely to discount them altogether. The scientist who believes in what he is doing, who has some kind of stake – personal or otherwise – in the successful outcome of his work, is likely to be convinced by them, to take them as justification and proof that he is not chasing wild geese. Herein lies the greatest danger of modern experimental methods.

Rhine, over a period of years, produced a large number of statistically significant results. He could quote 'odds against the results being produced by pure chance' ranging from 100–1 to 1,000,000–1. What this evidence added up to, in many minds, was 'conclusive proof' of ESP.

Many critics of Rhine, operating on the well-known Holmesian doctrine 'eliminate the impossible and whatever remains, however improbable, is the truth' and in the convinced belief that ESP was impossible, attacked him on the grounds that the findings must have resulted from cheating. This is quite unfair, and really quite unnecessary. One cannot deny evidence – one can only attempt to remain aware of the range of possibilities which might be indicated by that evidence. There has always been a tendency on both sides of the debate to minimise that range – to interpret the data in the light of personal convictions.

Quoted odds can be very misleading. The fact that the probability of achieving a particular result by pure chance is less than one in a hundred does not necessarily imply that it is 99 per cent certain that unknown forces are at work. Any particular sequence of events becomes unlikely when compared to the full range of possible events. Take, for instance, a sequence of five throws using a die. If the die came down 1 every time, one might begin to suspect that there was more than chance involved. After all, the probability of a die coming down 1 five times in a row by pure chance is only 1 in $6 \times 6 \times 6 \times 6 \times 6$ (1 in 7776 – statistically very significant). But suppose the die had come down any other number five times running – would that not have been equally significant? And suppose it had 'missed' once in the sequence – isn't four out of five still statistically significant? Suppose, too, that the sequence of numbers generated was the same backwards as forward (i.e. a palindrome) – wouldn't that be enough to attract our attention and make us calculate how unlikely such a thing must be? The odds against any particular palindromatic sequence are the same 1 in 7776, but the odds against the sequence being palindromic are a mere 1 in 36. And then again, suppose the sequence corresponded to my telephone number, or the height of Mount Everest, or … wouldn't that be not only 'significant' but positively uncanny?

The low probability of a particular event does *not* imply that something other than chance is at work. We draw such an inference not because of probability but because we see some kind of pattern in a sequence of events. There are all kinds of patterns we might look for and find in a sequence of events, and when we find them we are always ready to marvel – perhaps because, inevitably, we have faith in the fact that within the chaos of singular events there *are* patterns – simple guiding principles. That is, of course, the fundamental assumption of science, and the basis of religious faith. We believe in order, and randomness is an affront to our ideologies. It is not surprising that we find people with a passionate belief in systems for playing roulette, in lucky streaks and in the profound significance of coincidences. In a way, we might almost be said to possess a horror of randomness – a compulsion to find some kind of sense in the day-to-day sequence of events which make up our lives. Virtually all popular newspapers and magazines run horoscopes for the great majority of their readers who believe that there is 'something in it' – not being sure what 'it' is, but convinced that somehow there *must* be a pattern in life which makes it predictable, explains why things sometimes go wrong and sometimes go right.

There is, in a sense, a kind of schizophrenia about the scientist who is convinced that ESP research (and all the other subjects which occasionally are related to it by popularisers of scientific heresies in general) is all lies and madness. The faith which underlies these ideas is precisely the same as the faith which underlies his own conviction that science is absolute knowledge.

I remain unconvinced by Rhine's results for several reasons. One is that subjects who get high scores in Rhine's experiments almost invariably fail to sustain their high scoring. Rhine attributes this to the fact that subjects get tired, lose their talent and need 're-motivating'. There is probably a far simpler explanation for this lack of consistency, and there is an obvious danger in the policy of 're-motivating'. Making it worth a subject's while to produce results is not exactly conducive to an all-round attitude of scientific objectivity. In addition, there has been a tendency in Rhine's work to seek new correlations where the basic one failed to show up. If the series of symbols called by the subject failed to tally with the actual run, Rhine took to testing it against the sequence which was one step removed – i.e. correlating the subject's calls not with the card being shown but with the *next* card to be shown. This method doubled the number of significant results he obtained and provided him with as much evidence for the existence of 'precognition' as he was obtaining for the existence of 'telepathy'.

There is, on these bases, considerable ground to doubt Rhine's conclusions. Doubt, however, must work both ways. From a purely pragmatic point of view, there seem to me to be no grounds for believing in ESP – if no supposed talent can accomplish anything that moderate conjurors and illusionists cannot, then it really does not matter how the ends are accomplished. When a person who lays claim to extraordinary mental powers can accomplish something a competent stage magician cannot duplicate by trickery, then – and *only* then – one may admit that the idea of ESP might be useful. Until such evidence is forthcoming there is no reason why the existence of ESP should be accepted as a working assumption. At the same time, however, there is no reason why research should not continue uncondemned.

Though it remains a 'heretical' belief, resented by the majority of the scientific establishment, belief in ESP is supported by propagandist literature which mimics the literature of science very closely. Indeed, there are a good many fields of 'pseudoscientific' interest which have produced in recent years a rich literature which copies many of the strategies of scientific literature. Books on damned data and phenomena whose reality is not generally accepted record observational data and statistical data, they correlate and they deduce, they quote one another as sources. Even the form in which evidence is presented often mimics the form which scientists habitually use to present their data – for instance, they adopt the same euphemistic modesty, whereby 'I think' is rendered as 'Most authorities consider' or 'It is evident that' or even 'There is no doubt that'.

Surveys of this kind of activity, like Martin Gardner's *In the*

Name of Science and John Sladek's *The New Apocrypha*, which are intent on 'debunking' the beliefs which lie behind or spring from it, have no difficulty in finding instances of fake data being religiously quoted from source to source, often changing as they go, growing to fit the beliefs concerned. But this is not merely an indictment of the beliefs – it is also an illustration of the fallibility of the system. Scientific data, we are told, are sound because they can always be tested, always reproduced, and are checked by a host of workers within the scientific community. But is this so? How much checking does go on? If, by one means or another a misleading result or a downright lie *were* published in a *bona fide* scientific journal, how long would it take for it to be detected? In the meantime, how many people would quote and refer to and draw inferences from the datum if it happened to be relevant to their own beliefs or their own deductions from their experiments? How many experiments might be interpreted in the light of unreliable data so that the inferences drawn from them were not the same as the inferences which might otherwise have been drawn?

The fact is that the widely held attitude that scientific data are sacred and unchallengeable militates strongly against the possibility that inaccuracies can be quickly detected and filtered out from the network of scientific communication. Results are not checked many times by many workers – what scientist can justify his existence simply on the basis that he repeats other people's experiments to find out whether they are fools or liars? Published results are accepted – and believed – almost as a matter of course, *except when they conflict with whole theoretical contexts*. Then, of course, they are rejected out of hand.

The trouble is, of course, that theoretical contexts can only be competently revised and updated by acknowledging and recognising the importance of data which conflict with them. As things stand, the whole system is loaded *against* the acknowledgment and recognition of such data, while at the same time it is loaded in favour of data which confirm already-extant beliefs, whether such data are competent or not.

The faults of scientific heresies mirror the faults of the scientific establishment – in the perils of unscientific belief we can identify the perils of scientific belief also. The only 'tests' which personal beliefs have to undergo are pragmatic ones. Beliefs thrive because they are useful. But there is more than one kind of usefulness.

There is no need for belief in science – indeed, one might

almost say that there is no *room* for belief in science, for belief can only be a handicap to an open mind. One does not have to believe in an equation in order to use it – one need only recognise that it is the most useful conceptual tool available, quite dispensable if another one turns up which is even more useful. And yet scientists, like everyone else, do tend to be believers. If they do not believe in the particular aspects of science they tend to believe in a more general scientific ideology: they believe in the fundamental assumption of science, and they believe that the scientific method is the way to a proper understanding of the universe. How far this basic belief extends itself into the minutiae of scientific theory varies considerably between individuals.

It is becoming less and less easy to believe in the theories of science. This century has seen those theories become vastly more complex, and put much heavier demands upon the imagination and the intellect. As the ideas of science have been pushed beyond the limits of common sense their capacity to inspire belief has declined. This may mean little enough to the scientist, who is, in any case, committed, but to the layman it can mean a great deal. The nineteenth-century rationalists deliberately set out to destroy all the illusions contained within other structures of· belief, promising to deliver in their place the real, genuine, one-and-only key to the riddle of the universe. But the reality their twentieth-century successors have begun to discover is – in the minds of many – a substandard (i.e. less useful) substitute.

This, I think, is why we see today so many attempts to recover old beliefs in new packages, or even to design new ones useful in the same manner as the old, but mimicking the glamour of the intellectual armies which drove out the old. Their quest is for simplicity and common sense. The vast span of geological time is rendered irrelevant by the conviction that everything important that ever happened did so within the time-span of ancient history. The scale of cosmic events is reduced by the assumption that all events on Earth have causes understandable in a human context. The confusing world of the microcosm and the known forces which govern matter become irrelevant if we assume that the power of thought can overrule them all and that all matter is subject to the mind (if only we know how to control it). Despite the discovery of the macrocosm, we still want to believe that the sun exists to shine on Earth, and that if there are other worlds and other suns they only exist to provide eventual homes for man or homes for the extraterrestrial angels by reference to whom we can explain away (without leaving our armchair) all the mysteries of Earth.

The beliefs which are being promoted as alternatives to science are deliberately calculated to solve all our problems. Within their world-view, in fact, we *have* no problems of our own – everything which happens does so because of the bad aspect of Jupiter or because the devil is at work. We have no solutions of our own either, but that doesn't matter because either the flying saucer men will save us or our macrobiotic diet will make us perfect.

These beliefs are the measure of the disappointment with which the revelations of modern science are greeted. Bacon, Mercier, and the nineteenth-century rationalists earnestly believed that science would solve all our problems for us. It has not done so and it is no longer plausible to suggest that it will. It is a hopeless vanity to think that we can go out and discover the reality we would like to be a part of: a quest for an intellectual El Dorado no less mythical than the city of gold which Walter Raleigh sought.

The ability to believe is very useful – it is one of the faculties which allows us to disregard a good deal of the data which our senses feed to our minds, to sort out the relevant from the irrelevant, and engineer the faculty of choice which has played such an important role in the evolution of man and mind. But the ability to *resist* belief is also useful – perhaps more so now than ever before. Only by casting aside the shield of belief can we really hope to come to terms with the mysteries of modern science.

A Note in Conclusion

This has been a book about the mysteries of modern science. I have not attempted to compile a mere catalogue of enigmas, but have instead tried to show *why* modern science – because, rather than in spite of, its great intellectual triumphs – retains its mysteries.

Science began as one of several attempts to make sense of the universe. We are, in our manifold ways, still trying. Science has helped us transform ourselves and our world, but it has not done what it was initially intended to do. It has not provided us with THE ANSWER to 'the riddle of the universe'. It has not even given us hope that such an answer exists.

I have, perhaps, been rather unkind in my treatment of men like Pythagoras and Haeckel, whose faith that science would provide the answer led them into the belief that it already had. But in these men we can see the fallibility of the opinions of a great many others, whose ideas have not betrayed them quite so dramatically. Every confirmed believer should walk in fear of his own intellectual nemesis. There is an irrational number haunting every neat solution to the riddle.

I have spoken a great deal – perhaps in rather derisory terms – about 'common sense' and its limitations. I have tried, while retaining a great respect for the intellectual achievements of science and scientists, to demonstrate the hopeless vanity of the assumption that we can – aided by scientific belief or any alternative thereto – achieve a perfect understanding of the universe. My main concern has been to show that common sense cannot cope with the task, and that there will, of necessity, always be mysteries.

The march of science, particularly in the last century, has demonstrated quite clearly the fallibility of one of scientific faith's most trusted allies: Occam's razor. We have surely been shown that the Law of Parsimony is a seductive cheat. It leads us to the

263

simplest answers, not to the right ones. It may be ambitious to multiply hypothetical entities, but twentieth-century science has so often disclosed the fact that reality exceeds our ambitions. There are more real entities than we had dared hypothesise. The universe is complex, and we cannot make it simple by our own insistence on being simple-minded.

Isaac Newton, the man who synthesised the discoveries of the seventeenth-century physical scientists and astronomers into a new theory of the universe, and provided modern science with its first great edifice of organised knowledge, spoke of himself as follows: 'I do not know what I may appear to the world, but to myself I seem to have been only like a boy playing on the sea-shore, and diverting myself in now and then finding a smoother pebble or a prettier shell than ordinary, whilst the great ocean of truth lay all undiscovered before me.'

Today, we have a great collection of smoother pebbles, and a fine array of prettier shells – assemblies so great that it is easy to lose ourselves in examination and contemplation of their wonders. But we must realise, and never forget, that no matter how great the collection grows, the undiscovered ocean will still surround the shores of our imagination.

Index

Index